D0206375

Richard Taylor

NATURAL PHILOSOPHY THROUGH THE 18th CENTURY
AND
ALLIED TOPICS

NATURAL PHILOSOPHY THROUGH THE 18th CENTURY
AND
ALLIED TOPICS

Edited by ALLAN FERGUSON

"Nec aranearum sane textus ideo melior quia ex se fila gignunt, nec noster vilior quia ex alienis libamus ut apes."

JUST. LIPS. *Polit.* lib. i. cap. 1. Net.

TAYLOR & FRANCIS LTD
LONDON

ROWMAN AND LITTLEFIELD
TOTOWA, NEW JERSEY

1972

First published 1948 as a supplement to *The Philosophical Magazine* to mark the 150th anniversary of the founding of the journal.

This edition published 1972 by Taylor & Francis Ltd, London and Rowman and Littlefield, Totowa, New Jersey

Printed and bound in Great Britain by Taylor & Francis Ltd, 10–14 Macklin Street, London WC2B 5NF

Taylor & Francis ISBN 0 85066 055 6
Rowman and Littlefield ISBN 0 847471 092 8

"Meditationis est perscrutari occulta; contemplationis est admirari perspicua.... Admiratio generat quæstionem, quæstio investigationem, investigatio inventionem."—
Hugo de S. Victore.

"Cur spirent venti, cur terra dehiscat,
Cur mare turgescat, pelago cur tantus amaror,
Cur caput obscura Phœbus furrugine condat,
Quid toties diros cogat flagrare cometas,
Quid pariat nubes, veniant cur fulmina cœlo,
Quo micet igne Iris, superos quis conciat orbes
Tam vario motu."

J. B. Pinelli ad Mazonium

Preface

The Philosophical Magazine is an independently owned journal; it is not the organ of any institution or scientific body. As a scientific journal it is therefore in a unique position.

The writer in his imagination can weave an historical romance through the years of the existence of the Magazine and of the firm of Taylor & Francis, its proprietors and printers. It is not, however, his intention to lapse to any length in that direction, as the reader will have ample opportunity to build a picture from the interesting articles contributed to this Commemoration number of one hundred and fifty years in the life of the Magazine.

As Chairman of the Board of Directors of the firm it is a privilege to be given the task of writing this note of introduction to the illuminating articles which follow. The writer would like to take this opportunity of expressing his admiration for the courage of the Executive Editor who has successfully steered the Magazine through the years of the war, and, despite physical disabilities, has planned and produced this Commemoration number. He has succeeded in obtaining most interesting articles from able writers whose names are familiar to the scientific world and who hold positions of eminence in the field of science.

Few readers can be aware of the character of the home of *The Philosophical Magazine*, and some may be unmindful of the many fond parents who have attended to its welfare throughout the years of its growth to maturity. During those years it has achieved an international reputation as a scientific journal in the realm of physical and mathematical science. Let the reader allow his imagination to recall the history of the times of the notorious Bishop Bonner who was resident in a house forming part of the premises where the Magazine is still printed, and let him visualize the many editors who have ascended the seventeenth century staircase and then allow his imagination free play as he wanders through the rooms, three of which have ornate ceilings attributed to Sir Christopher Wren. He cannot fail to be impressed by the surroundings of the building, the walls showing signs of their age and of additional scars from the recent war. Further, he will see among the workers many who have given loyal service to the firm from their early youth.

Editors and workers know too well the difficult times through which the Magazine and the firm have passed and yet the firm has succeeded in producing journals and other publications for many academic and scientific institutions among whom they have earned a high reputation as scientific printers and publishers.

The Philosophical Magazine has indeed passed through the trials and viscissitudes of life but the editors feel with confidence that the past achievements of the Magazine justify their greater confidence in its continued success.

The Editorial Board—Sir Lawrence Bragg, Sir George Thomson and Dr. Allan Ferguson—has been strengthened by the addition of Professor N. F. Mott, who takes on the duties of Editor. The Board of Directors wish them every success in the journey towards the 200th anniversary of the foundation of the Magazine.

E. J. BURDON

Introduction to the 1972 Edition

by

Sir Nevill Mott, F.R.S.
Chairman of Taylor & Francis Ltd

THIS collection of articles on science and engineering was published first in 1948 to celebrate the hundred and fifty years in which *The Philosophical Magazine* has served the scientific community. The interest in the history of science now shown in schools and universities has led Taylor & Francis to issue this reprint. We believe that many of these articles would form an admirable introduction to more detailed study. The title-page alone of the first number, describing *The Philosophical Magazine* as comprehending the various branches of science, the liberal and fine arts, agriculture, manufactures and commerce, shows, if nothing else, how the meaning of the word philosophy has changed. Successive editors have written to many publishers, returning books sent for review on what we now call philosophy and have explained that though we value tradition and therefore keep our old title, we publish mainly solid state physics. But if this book is to be used, as I hope, in schools and colleges of education where people are thinking about how to teach the history of science, one of the most useful articles may be that by Sherwood Taylor on the teaching of science at the end of the eighteenth century. Readers will note how little the education of famous scientists was related to their future achievements. Avogadro was a lawyer, and Sedgwick, the famous Cambridge geologist, was a mathematician and knew little or no geology until he was appointed to his chair. Neither Ampère nor Faraday had any formal higher education, and it will surprise no-one who has plucked the wires of a Wheatstone bridge in his school laboratory to learn that Wheatstone was a musical instrument maker. Perhaps this was just as well, if the assessment of teaching at the end of the eighteenth century in Captain Smith's article on engineering and invention in the eighteenth century is fact. A contemporary writes: "I have often with much sorrow bewailed the misfortunes of the children of Great Britain, when I consider the ignorance and undiscerning of the generality of schoolmasters. . . . Many of these stupid tyrants exercise their cruelty without any manner of distinction of the capacities of children, or the intentions of parents in their behalf". Few of the pioneers at that period owed much to their educators.

We have come a long way from then. Indeed, Dr. Allan Ferguson's article serves as a reminder of the further progress that science education has made since this itself was written. He lists some of the important papers that appeared during the nineteenth century and up to the 1930s, and the majority of these are easy to follow. An education which aims at developing the capacity for scientific thinking must work close to the original records of scientific thought, and the past volumes of *The Philosophical Magazine* over this period are a rich store of source material for the science teachers of today. This is no fanciful claim; the exploratory Nuffield courses for teachers in the early days of the project brought them into direct contact with the papers of Bragg and Moseley and Rutherford, and Perrin's famous paper on The Brownian Movement (later reprinted separately) must have inspired the way in which young pupils nowadays first come to terms with molecules. Dr. Ferguson says that the volumes commemorated by this publication " display the development of scientific method and achievement the more strikingly because unconsciously ".

In the new climate of scientific education, their contribution to the service of the scientific community will extend far into the future.

N. F. Mott

Contents

THE

PHILOSOPHICAL MAGAZINE.

COMPREHENDING

THE VARIOUS BRANCHES OF SCIENCE,

THE LIBERAL AND FINE ARTS,

AGRICULTURE, MANUFACTURES,

AND

COMMERCE.

BY ALEXANDER TILLOCH,

MEMBER OF THE LONDON PHILOSOPHICAL SOCIETY.

"Nec aranearum ſane textus ideo melior, quia ex ſe fila gignunt. Nec noſter vilior quia ex alienis libamus ut apes." JUST. LIPS. *Monit. Polit.* lib. i. cap. 1.

VOL. I.

LONDON:

Printed for ALEX. TILLOCH: And sold by Messrs. RICHARDSON, Cornhill; CADELL and DAVIES, Strand; DEBRETT, Piccadilly; MURRAY and HIGHLEY, No. 32, Fleet-street; SYMONDS, Paternoster Row; BELL, No. 148, Oxford-street; VERNOR and HOOD, Poultry; HARDING, No. 36, St. James's-street; J. REMNANT, High-street, St. Giles's; and W. REMNANT, Hamburgh.

Reproduction of title-page to first volume.

THE PHILOSOPHICAL MAGAZINE

By ALLAN FERGUSON, M.A., D.Sc.
AND
JOHN FERGUSON, M.A., B.D.

THERE is an obvious danger in associating great movements with the turn of a century, but the years about 1800 are associated with scientific and social events of the first magnitude. Volta in 1800 was to announce his invention of the voltaic pile and the " crown of cups " ; these systems were the first examples of practicable forms of battery producing a continuous current. The discovery was used almost at once by Nicholson and Carlisle to decompose water by electrolysis. In 1799, Benjamin Thompson, Count Rumford, founded the Royal Institution of Great Britain. About the same time that versatile genius, Thomas Young, was carrying out those experiments on the interference of light which were to rehabilitate the wave theory, though their influence was to be delayed by a spate of abuse from Brougham (" the feeble lucubrations of this author in which we have searched without success for some traces of learning, acuteness and ingenuity "). In 1798 Cavendish published the result of his classic torsion balance experiment on the Earth's mean density—a result in which that most accurate of experimenters made a slip in calculating his final mean, and so put on record his final result as 5·48 instead of 5·448.

The social aspect of England was changing rapidly, and the era of steam was looming ahead. In 1812 the *Comet*, Henry Bell's steamboat, was to be seen making its smoky progress along the Clyde ; the canal and road system of the country was taking shape, and the stage was set for the vast industrial expansion of the nineteenth century, an expansion bringing with it problems of affluence and poverty for which it is taxing all the resources of the twentieth century to find a solution. Interest in matters of science, pure and applied, was spreading, and the audience to which such topics appealed was increasing in numbers and intelligence. It was in circumstances such as these that Alexander Tilloch founded the Philosophical Magazine, the first number bearing the date June 1798.

Alexander Tilloch was a prominent figure in the contemporary world of invention and printing. He was born at Glasgow in 1759, the son of John Tulloch, a tobacco merchant and magistrate. He was educated at the University of Glasgow. About the year 1787 he changed his surname from Tulloch to Tilloch. He became interested in printing and independently rediscovered the process of stereotyping, for which, in 1784 he took out a patent jointly with Andrew Foulis, the printer. Part of the story may be read in an article in the Philosophical Magazine (vol. x. p. 267, 1801) over Tilloch's initials. Tilloch migrated to London in 1787, where, among various printing adventures which included an unsuccessful attempt to persuade the Bank of England to adopt a method of printing notes which would make forgery impossible, he founded the Philosophical Magazine. In later life Tilloch became deeply interested in the subject of prophecy and joined the Sandemanian sect, a sect of which Michael Faraday was an elder. Tilloch died in 1825 in Islington, near London.

But, in the opening years of the nineteenth century, a new name appeared in the printing world—the name of one who was to contribute in no small degree to the setting of a high standard of printing, especially of scientific works, and to show that a critical appreciation of the latest advances in science was not incompatible with a generous knowledge of and an enthusiasm for the humanities. Richard Taylor (1781–1858) was the second son of John Taylor, wool-comber and hymn-writer, and was born at Norwich. He was educated at a private school in his native city and, at the suggestion of Sir James Edward Smith (the founder of the Linnean Society), was apprenticed to an eminent printer of Chancery Lane, London, Jonas Davis by name, whose press had been active in London since the middle of the eighteenth century. Richard Taylor's adventures in printing may be followed in the pages of the Philosophical Magazine where appear, on title-page or as colophon, the name of the printer and the place of his press. The data are collected in appendix I., where it will be seen that in 1800 the firm Jonas Davis became Davis, Taylor & Wilks, and in 1801 the name of Davis disappears. Richard Taylor set up the Black Horse Court press in partnership with his father; Arthur Taylor, who appears as a partner from 1814 to 1823, was a younger brother of Richard, and John Edward Taylor, partner from 1837 to 1851, was a nephew. In 1852, Dr. William Francis, of whom we shall have more to say later, became a partner in the firm.

Richard Taylor was a man of wide learning and of liberal views. He was a competent classical scholar, and had made a special study of mediæval and renaissance writers in Latin and in Italian. He was a student of the French, Flemish and Anglo-Saxon languages and literatures, and he published many works dealing with Anglo-Saxon literature. In 1810 Taylor was appointed under-secretary to the Linnean Society, a position which he filled with distinction for nearly half a century. In 1822 he became joint-editor with Tilloch of the Philosophical Magazine. In 1838 he initiated, with the help of William Francis, a journal which in 1841 became the Annals and Magazine of Natural History, which has continued to the present day, and is now chiefly concerned with systematics. His feeling for the needs of the times is well shown by his publication of Taylor's Scientific Memoirs, a series of translations of important scientific papers, which appeared in five stout volumes between 1837 and 1852.

Taylor was a good citizen; he played a large part in the foundation of the City of London School and of University College, and he did much for the corporation library.

The quotations from mediæval authors which are found on the obverse and reverse of the title-pages of each volume of the Philosophical Magazine are characteristic of an age of scholarship rather than specialization; an age in which William Thomson, the 22-year-old professor of natural philosophy in the University of Glasgow, could compose and read in Latin his inaugural lecture, " De Caloris Distributione per Terræ Corpus," of which feat his friend and successor Andrew Gray remarks " it is unlikely that the form of the inaugural dissertation cost him much more trouble than the matter." Indeed, this feat, which would not be regarded by his contemporaries as anything out of the ordinary, affords material for an interesting commentary on Johnson's gibe against the Scots: "Their learning is like bread in a besieged town; every man gets a little, but no man gets a full meal."

The quotation from Justus Lipsius appears on the title-page of the first volume, and although Justus Lipsius was one of Richard Taylor's favourite authors in later years, it is barely possible that he had a hand in the selection of this passage *.

Richard Taylor broke down in health in 1851 and Dr. William Francis, who took as large a share in moulding the fortunes of the Philosophical Magazine during the second half of the 19th century as Richard Taylor had taken in the first half, became a partner in the firm and took over Taylor's share of the editorial work. Richard Taylor died at Richmond in 1858.

Dr. William Francis was born in London in 1817 and was educated at University College School and St. Omer. After a period of study at University College, London, he went to Berlin and later to Giessen, where he worked on the salts of molybdenum under Liebig, and graduated Ph.D. in 1842. He had been apprenticed to Richard Taylor and did good service to the Philosophical Magazine by translating important foreign papers for that journal before his name appeared on the editorial board in 1851. His name remained there till his death in 1904. During this long period, his wide knowledge, great ability, and his capacity for friendship had no small share in forwarding the influence of the journal.

Such, in brief, were the characters of the founders of the Philosophical Magazine and their successors. What of the journal itself? It began its career as a miscellany of scientific information and research, and its aims are set out in characteristic language in the preface to the first volume, which preface is reprinted here as appendix IV., retaining even a delightful printer's error, which has been indicated by (*sic* !).

" New or curious." A rummage amongst the early volumes discloses much that is both new and curious. Volume 9, for example, contains an article on " A Singular Case of Dropsy " and recounts the story of an unfortunate boy of 12 who was tapped twice and on each occasion

---

* Translations of the passages on the obverse and reverse of the title-page appear in appendix II.

A note on the house-symbol appears in appendix III.

" 7–8 quarts of milky fragrant chyle " were withdrawn. Volume 11 contains a dissertation on the problem " Whether mixed metals can be distinguished by the Smell " ; and volume 14 has a paper on a " New Theory of the Constitution of Mixed Gases elucidated by J. Dalton, Esq." Volume 19 has a paper by C. Gordon, M.D. on the Antiquity of the Gealic (*sic*) language in which the learned author links the Gealic *Precept* with the Hebrew *Decalogue*, and proves to his own satisfaction that Gealic, not Hebrew, is the primitive language of mankind. The more serious scientific papers had, in many instances, appeared elsewhere before publication in the Philosophical Magazine. Davy's Bakerian lectures, many of which were reprinted in the Philosophical Magazine, are cases in point. Some of the titles of important or curious articles in the earlier volumes are given in appendix V.

As the century wore on the Philosophical Magazine tended more and more to become a journal of research. It had absorbed two journals—the Journal of Natural Philosophy, Chemistry and the Arts (1797–1814) commonly known as Nicholson's Journal, and Thomas Thomson's monthly scientific journal the Annals of Philosophy (1813–1827), and in the year of incorporation of Thomson's Annals (1827) the name of an external advisor, Richard Phillips, appears on the editorial board. In 1832 Sir David Brewster's name appears ; the subsequent changes in the board of editors are set out in appendix VI.

A journal of research necessarily admits papers of very different degrees of importance. There is much honest, interesting and accurate work which merits publication, but of which it cannot be said that its rejection would appreciably retard the advancement of science, such as happened in the rejection by the Royal Society of Waterston's fundamental paper on the kinetic theory of gases. The refusal to publish this paper delayed the development of the kinetic theory by ten or fifteen years. It is unfortunate that Waterston did not select the Philosophical Magazine as his medium for publication, as he did for many of his investigations. H. A. Rowland, one of the outstanding experimental physicists of the 19th century, could find no American journal which would accept his first scientific paper, sent it to Clerk Maxwell, who at once communicated it to the Philosophical Magazine where it appeared in 1873.

The Philosophical Magazine has indeed had its share of important papers. In 1851 Sir William (Rowan) Hamilton contributed a long series of papers on extensions of the notion of quaternions. William Thomson (Lord Kelvin) contributed many papers. The paper by Michelson and Morley on the relative motion of matter and ether appeared in the volume for 1887.

Sir Joseph Thomson was ever a faithful friend of the Philosophical Magazine. In his bibliography of 241 papers, no fewer than 85, including the first (1876) and the last (1939) appeared in the Philosophical Magazine, and the 85 include some of fundamental interest.

Rutherford's epoch-making papers on the artificial transmutation of the elements—a series which marks the close of his Manchester period—

appeared in the Philosophical Magazine for 1919. Papers involving significant advances in atomic physics have also been contributed by Bohr, Moseley, Chadwick and Millikan.

The years of the second world war brought their special problems. It became difficult to find papers to publish and to find paper on which to print them, but the journal continued to appear though each number was reduced from 8 sheets to 4½ sheets. And the problem of the paper supply has not become much easier since the end of the war. Nevertheless, in the last decade the journal has contained papers of outstanding interest which include Milne's paper on kinematical relativity, radio papers from Sir Edward Appleton, and papers from the research laboratories of the Scandinavian countries.

"Wonder breeds curiosity, curiosity investigation, investigation discovery," and from the relation of things new and curious has sprung the systematic application of the powers of the mind to the understanding of nature, and the collection and collation of the results thus obtained in orderly exposition. The 150 years of the Philosophical Magazine display the development of scientific method and achievement the more strikingly because unconsciously. In that development the journal has itself played no small part and continues to-day to provide a vehicle whereby may be spread abroad the products of the devoted scholarship of scientific men.

APPENDIX I.

Printers of the Philosophical Magazine.

1798. Volumes 1, 2 & 3.—No printer's impression.
Volumes 4 & 5.—Davis, Chancery Lane.
1800. Volume 6.—Davis, Taylor & Wilks.
1801. Volume 11.—Wilks & Taylor, Chancery Lane.
1802. Volume 14.—Taylor & Wilks (but not so Volumes 13 & 15).
1803. Volume 16.—Taylor, Black Horse Court, Fleet Street.
[Volume 17.—J. Taylor.]
1804. Volume 19.—R. Taylor, Black Horse Court, Fleet Street.
1805. Volume 20.—R. Taylor & Co., Black Horse Court, Fleet Street.
Volume 22.—R. Taylor & Co., 38 Shoe Lane, Fleet Street.
1814. Volume 43.—Richard & Arthur Taylor, Shoe Lane.
1823. Volume 61.—Richard Taylor, Shoe Lane.
1827. Volume 2, *Second Series.*—Richard Taylor, Red Lion Court, Fleet Street.
1837. Volume 10.—Richard & John E. Taylor, Red Lion Court.
1851. Volume 1, *Fourth Series.*—Richard Taylor, Red Lion Court.
1852. Volume 3.—Taylor & Francis, Red Lion Court.

APPENDIX II.

Translation of the passages appearing on the obverse and reverse of the title-pages of each volume of the Philosophical Magazine.

Justus Lipsius "Nec aranearum . . . . "

The way the spiders weave, you see, is none the better because they produce the threads from their own body, nor is ours the worse because like bees we cull from the work of others.

Hugo de S. Victore " Meditationis est . . . . "

Meditation means searching out that which is hidden, contemplation regarding with wonder that which is not. Wonder breeds curiosity, curiosity investigation, investigation discovery.

J. B. Pinelli " Cur spirent venti . . . . "

Why the winds blow, why fissures appear in the earth, why the sea swells, why the ocean is so salt, why the sun hides his head in ruddy darkness, what makes the comets blaze forth so often and so terribly, what gives rise to clouds, why thunderbolts appear in the sky, with what fire the rainbow flashes, who summons together the heavenly bodies in such varied courses.

The passage from Justus Lipsius appeared on the title-page of Volume 1 (1798) ; the passage from Hugo de S. Victore appeared on the reverse of the title-page in 1845 and the passage from Pinelli was added in 1849.

APPENDIX III.

The house-symbol, the lighted lamp with its motto " Alere flammam " appears first in 1823 on the reverse of the title-page. In 1835 it is transferred to the end of the volume, and in 1883 it reappears on the reverse of the title-page.

APPENDIX IV.

Preface to Volume 1.

Philosophical Magazine.

1 1798

Preface

Having concluded our first Volume, we would be deficient in gratitude did we not return thinks (*sic* !) to the public, in general, for the favourable reception our labours have experienced ; and to those Scientific Gentlemen, in particular, who have assisted us with Communications, as well as Hints respecting the future conducting of the Work.

As the grand Object of it is to diffuse Philosophical Knowledge among every Class of Society, and to give the Public as early an Account as possible of every thing new or curious in the scientific World, both at Home and on the Continent, we flatter ourselves with the hope that the same liberal Patronage we have hitherto experienced will be continued ; and that Scientific Men will afford us that Support and Assistance which they may think our Attempt entitled to. Whatever may be our future Success, no Exertions shall be wanting on our part to render the Work useful to Society, and especially to the Arts and Manufacturers of Great Britain which, as is well known, have been much improved by the great Progress that has lately been made in various Branches of the Philosophical Sciences.

APPENDIX V.

Some titles of articles from the earlier volumes.

Volume 9.—" History of Astronomy for the year 1800," by Jerome de Lalande.
" Singular Case of Dropsy."

Volume 11.—" A dissertation on the History of Sugar," by Professor Beekmann.
" Observations on the Onager of the Antients, or the real Wild Ass," by Professor Pallas.

Volume 14.—Several articles by Kirwan.
" On Painting," by Mr. E. Dayes, Painter.

Volume 19.—" A Report of the State of His Majesty's Flock of Fine-woolled Spanish Sheep, for the year ending Michaelmas 1803," by Sir Joseph Banks, P.R.S.

Volume 23.—" Observations on the singular figure of the planet Saturn," by William Herschel, LL.D., F.R.S.
" Abstract of Observations on a diurnal Variation of the barometer between the Tropics," by J. Horsburgh, Esq., in a letter to Henry Cavendish, Esq., F.R.S.

Volume 32.—Davy's Bakerian lecture on Alkalis.
Memoirs of the late Erasmus Darwin, M.D.

Volume 33.—" Analysis of the Mécanique Céleste of M. La Place," by M. Biot.

Volume 36.—" On Crystallography," by M. Haüy. Translated from the last Paris edition of his ' Traité de Minéralogie.

Volume 38.—Memorandum on the subject of Lord Elgin's Pursuits in Greece.
" Some speculations on the Nature of Instinct," by Arthur Mower.
The reports of Mr. William Smith and Mr. Edward Martin to the Bristol and Taunton Canal Company on the State of the Collieries at and near Nailsea, in Somerset.

Volume 43.—" Experiments tending to prove that neither Sir Isaac Newton, Herschel nor any other Person, ever decomposed incident or impingent Light into the prismatic Colours," by Joseph Reade, M.D.
" On the Capacity of Heat or Calorific Power of Various Liquids," by Benjamin, Count Rumford.

Volume 51.—" Whether Music is necessary to the Orator—to what extent and how most readily attainable ? " by Henry Upington, Esq.
" On the Case of Miss Margaret McAvoy."
" On the Atomic Theory," by William Higgins, Esq.

Volume 54.—Report by the Select Committee of the House of Commons on Mr. Telford's Plan for hanging an Iron Bridge across the Menai Straits.

Volume 62.—Report of T. Telford, Esq., on the Effects which will be produced on the Thames by the rebuilding of London Bridge.
" On Fluid Chlorine," by Mr. Faraday.
" On the Condensation of several Gases into Liquids," by Mr. Faraday.

Volume 65.—" Description of an improved Cross for Land-Surveyors," by by Mr. Isaac Newton.
" An Experimental Inquiry into the Nature of the radiant heating Effects from terrestrial Sources," by Baden Powell, M.A.
" On the Osteology of Reptiles, and on the Geological Position of their fossil remains," by Baron Cuvier.

Volume 66.—" On new compounds of Carbon and Hydrogen," by M. Faraday, F.R.S.

New Series.

Volume 10.—Professor Whewell on Isomorphism.

Volume 11.—First report on British Association for the Advancement of Science.
Various papers by Sir D. Brewster including one on " An Account of a Curious Chinese Mirror, which reflects from its polished face the figures embossed upon its back."

APPENDIX VI.

List of Editors.

Table I.

Alexander Tilloch	1798–1825
Richard Taylor	1822–1858
Richard Phillips	1827–1851
David Brewster	1832–1868
Robert Kane	1840–1889
William Francis	1851–1904
John Tyndall	1854–1863
Augustus Matthiessen	1869–1870
William Thomson	1871–1907
George Francis Fitzgerald	1890–1901
John Joly	1901–1934
William Francis (Jr.)	1904–1932
Oliver Joseph Lodge	1911–1940
Joseph John Thomson	1911–1940
George Carey Foster	1911–1919
Richard Taunton Francis	1921–1931
Alfred W. Porter	1931–1939
John R. Airey	1932–1937
Allan Ferguson..............	1937–
Lawrence Bragg	1941–
G. P. Thomson	1941–

Table II.

1798. Alexander Tilloch.
1822. Alexander Tilloch and Richard Taylor.
1825. Richard Taylor.
1827. Richard Taylor and Richard Phillips.
1832. Sir David Brewster, Richard Taylor and Richard Phillips.

[From 1835 to 1844, a note appears on the reverse of the title-page :– " The Conductors . . . . beg to acknowledge the editorial assistance rendered them by their friend Mr. Edward William Brayley, F.L. & G.S., Librarian to the London Institution ; which commenced in the year 1823, in Vol. 61 of the First Series, and has been regularly continued from that time.]

1840. Sir David Brewster, Richard Taylor, Richard Phillips and Robert Kane.
1851 (Vol. 1). *Add* William Francis.
(Vol. 2). *Omit* Richard Phillips.

1854. *Add* John Tyndall.
1859. *Omit* Richard Taylor.
1864. *Omit* John Tyndall.
1868. Robert Kane and William Francis (*omit* Brewster).
1869. *Add* Augustus Matthiessen.
1870. *Omit* Augustus Matthiessen.
1871. Robert Kane, William Thomson and William Francis.
1890. *Add* George Francis Fitzgerald. *Omit* Robert Kane.
1901. *Add* John Joly. *Omit* G. F. Fitzgerald.
1904. William Francis (Jr.) replaces his father.
1908. *Omit* Lord Kelvin (William Thomson).
1911. Oliver Lodge, J. J. Thomson, John Joly, George Carey Foster and William Francis.
1919. *Omit* G. Carey Foster.
1921. *Add* Richard Taunton Francis.
1931. *Add* A. W. Porter. *Omit* Richard Taunton Francis.
1933. *Omit* William Francis. *Add* John R. Airey.
1934. *Omit* John Joly.
1937. *Add* Allan Ferguson.
1938. *Omit* J. R. Airey.
1939. *Omit* A. W. Porter.

Leaving before end of 1939 until end of 1940 :—Sir Oliver J. Lodge, Sir Joseph J. Thomson and Allan Ferguson.

1940. Sir Lawrence Bragg, Sir George Thomson and Allan Ferguson.
1948. *Add* N. F. Mott.

ASTRONOMY THROUGH THE EIGHTEENTH CENTURY

By Sir H. SPENCER-JONES, F.R.S.

Up to the beginning of the eighteenth century astronomy had been concerned primarily with a single great problem, the plan of the solar system. Observations had been largely restricted to the determination of the positions of the Sun, Moon, planets, and fixed stars, which provided the material for the investigation of this problem. The formulation by Newton of the principle of universal gravitation, enabling him to account for the motions of the planets according to Kepler's three laws and for the more important inequalities in the motion of the Moon, had at length provided in broad outline the solution of this problem. A beginning had been made in the application of the telescope to astronomical observation; poor in quality though the early telescopes were, the first-fruits of their use had been well worth gathering. The discovery of spots on the Sun and of the Sun's rotation, of the phases of Venus, of mountains and other formations on the Moon, and of the four major satellites of Jupiter had been made by Galileo with his primitive telescope. The grinding and polishing of lenses of larger aperture was undertaken by Hevelius in Dantzig, by Christiaan and Constantyn Huygens at the Hague, and by Campani in Rome. Hevelius, by his study of the surface of the Moon, had laid the foundations of selenography; Christiaan Huygens had recognized the ring structure of the appendages of Saturn, which had puzzled Galileo, and had discovered Titan, the brightest satellite of Saturn; J. D. Cassini, using lenses made by Campani, had discovered the rotation of Mars, four satellites of Saturn (Iapetus, Rhea, Dione, and Tethys) and the division in Saturn's ring system, known as the Cassini division. But the large aberrations of the simple lenses of that time necessitated greater and greater focal lengths as apertures were increased, so that telescopes became unwieldy and awkward to manage. The difficulties of observation were so great that the long focus telescopes had soon fallen into disuse; many years elapsed before the satellites of Saturn which Cassini had discovered were once again seen.

During the greater part of the eighteenth century, astronomical observations continued to be largely concentrated on measurements of position. The application of telescopic sights, improvements in the design and construction of astronomical instruments, and refinements in the methods and technique of observation enabled a progressive increase in accuracy to be attained. Hevelius, who died in 1687, was the last astronomer of importance to use open sights for observations of position. It is surprising that such an assiduous observer, who had employed the telescope to such good effect in his observations of the Moon's surface, should have steadfastly refused to use it on his quadrants and sextants.

The observations of Tycho Brahe had been sufficiently accurate to enable Kepler to deduce from them his laws of planetary motion. Though the observations of Hevelius were more accurate than Tycho's, they proved to be of little scientific value. The time had passed when any important discovery could be made by using measures of position with an accuracy limited to about one minute of arc.

The accuracy of astronomical observations had in fact by the end of the seventeenth century lagged behind what was required for essentially practical purposes. A reliable solution of the problem of determining the position of a ship at sea had become a matter of great practical importance. When Charles II appointed a committee to consider a proposal to determine the longitude at sea by a method which was essentially that of lunar distances, John Flamsteed reported that in theory the longitude could be determined in this way; but, in practice, because the positions of the stars in the best available star catalogue could be in error by as much as 5′ and the position of the Moon derived from the current lunar tables might be in error by one-third of a degree, the derived longitude could be as much as 300 leagues in error. Flamsteed saw clearly what was the most important need of astronomy at that time; it was the systematic observation, continued for many years and with the highest attainable accuracy, of the positions of the fixed stars, the Sun, the Moon, and the planets, so that the positions of the stars could be improved and data provided from which tables of the motions of the Moon and other bodies could be constructed. He recommended that an observatory should be built for this purpose and provided with instruments and with persons, skilled in mathematics and astronomy, to make the observations. The outcome was the foundation of the Royal Observatory at Greenwich by Charles II in 1675 and the appointment of Flamsteed as the first Astronomer Royal. Flamsteed laboured for 44 years at Greenwich, setting the tradition for systematic and accurate observation which has characterized the Greenwich Observatory throughout its long history. He introduced new methods of observation, some of which have become standard practice since his time, and he established the fundamental points of practical astronomy upon a new basis. His observations exceeded in accuracy those of his predecessors or contemporaries and are, in fact, the earliest observations from which the phenomenon of aberration is clearly deducible. His great catalogue of the positions of nearly 3000 stars was an important contribution to the astronomy of his day.

The Paris Observatory, commenced in 1667 and completed in 1671, a few years prior to the Greenwich Observatory, had a brilliant start under its first director, G. D. Cassini, whose observations with long focus telescopes have already been mentioned. The observations by Picard in Paris and by Richer in Cayenne of Mars at its close approach in 1672 provided the best determination up to that time of the Sun's distance. It was found that a pendulum of given length beat more slowly at Cayenne than at Paris, showing that the intensity of gravity was less near the equator than in higher latitudes and suggesting that the earth was not a perfect sphere. Roemer, a young Danish astronomer, who in 1671 had been brought to Paris by Picard, proved in 1675, from the study of the times of eclipses of Jupiter's satellites, that light travels with a finite velocity, which he was able roughly to estimate. But in spite of this auspicious beginning, the Paris Observatory produced little of real importance during the greater part of the eighteenth century. No special duties were assigned to the Director and no definite plan of observation was projected. Thus it came about that the greater part of the best observing work of the eighteenth century was done in Great Britain.

We have mentioned that Newton had provided in broad outline the solution of the problem of the motions of the solar system. He left to his successors the task of accounting in detail for these motions, a task of incredible difficulty, which in effect created a new department of astronomy, gravitational astronomy. But though the Newtonian theory of universal gravitation was slower in gaining acceptance on the Continent, and particularly in France, where Descartes' theory of vortices held sway, than it was in England, Newton's own countrymen contributed little during the eighteenth century to the development of gravitational theory. It was to a group of brilliant continental mathematicians and geometers that the foundations of celestial mechanics were due. Newton had written the ' Principia ' almost entirely in the language of geometry ; though he had doubtless first solved many of the problems by his method of fluxions, he translated the results into geometry. He probably preferred not to employ in the ' Principia ' mathematical methods that were not familiar. Excessive reverence for Newton on the part of his countrymen, together with the estrangement between British and Continental mathematicians, which developed from the quarrel between Newton and Leibnitz as to their respective claims to the invention of what Newton termed fluxions and Leibnitz termed the differential method, prevented British mathematicians from using the powerful analytical methods that were being developed on the Continent. British mathematicians were in consequence almost isolated from the main trend of mathematical development during the eighteenth century. The great names in the development of gravitational astronomy, which began about 50 years after the publication of the first edition of the ' Principia ' and continued to the end of the eighteenth century, were Euler (1707–83), Clairaut (1713–65), D'Alembert (1717–83), Lagrange (1736–83), and Laplace (1749–1827). This development was, however, closely interlinked with the improvements in the precision of observation, whereby a more accurate control over the results of theory was provided.

Improvements in the construction of clocks had by the last decade of the seventeenth century made it practicable to determine the right ascensions of celestial bodies by observations of the times of their transits across the meridian, using a telescope fixed in the meridian. Roemer had realized, in order that a telescope should move as nearly as possible in the meridian, that it was much better to support it from the two ends of an axis rather than to mount it, as in the quadrant, from a short pivot. In 1689 Roemer invented an instrument which he called the *machina domestica*, which was the first transit instrument. The construction first met with approval in England and in 1722 Halley installed the first transit instrument to be used at Greenwich. It became standard practice to use the transit instrument for the observation of right ascensions and the mural quadrant for the observation of declinations. Roemer in 1704 had invented another instrument, called the *rota meridiana*, which was the first transit circle and could be used to determine both right ascensions and declinations. But the transit circle seems to have been forgotten and did not reappear until it was revived by Ramsden at the end of the eighteenth century.

Improvements in the design and construction of positional instruments were due mainly to the celebrated English makers Graham, Bird, Ramsden, and Troughton, who supplied instruments to many of the principal observatories. Improved methods of dividing arcs were developed and by the end of the century the errors of graduation had been lowered to about 2″ or 3″. The magnificent series of observations made by Bradley at Greenwich, particularly those made between 1750 and 1762, amounting to about 60,000, formed the most important contribution of the century to meridian astronomy. They showed a considerable advance in accuracy over those of Flamsteed, which represented the best that had been previously made : they are in fact the earliest observations which are of sufficient accuracy to be combined with modern observations for the investigation of the proper motions of the stars. The accuracy was due in some measure to the good construction of Bradley's instruments, made by Graham and Bird, but even more to his skill in observation, to the care with which he controlled the adjustments of the instruments, and also to his improved tables of atmospheric refraction, in which account was for the first time taken of the effects of temperature and of barometric height.

Various attempts to determine stellar distances by measurement of annual parallax proved unsuccessful. It was not until 1838 that the distance of a star was first determined. Bradley was one of the astronomers who had tackled the problem, by the observation of the zenith distances of the star γ Draconis, using a $12\frac{1}{2}$-foot zenith sector by Graham ; the attempt, though unsuccessful, led Bradley to the discovery in 1729 of the aberration of light, thereby incidentally providing the first observational proof of the motion of the earth round the Sun. Some small residual effects, which Bradley suspected to be due to the Moon, induced him to continue his observations through the period of revolution of the Moon's nodes (about 19 years), and enabled him to announce in 1748 the discovery of the nutation of the earth's axis. Bradley gave the correct explanation of nutation as a variation in the apparent annual precession caused by the variation in the gravitational action of the Sun and Moon on the equatorial parts of the earth.

Newton had inferred from theoretical considerations that the earth was spheroidal, being flattened towards the poles. Richer's pendulum experiments at Cayenne had given evidence of a departure from the spherical form. But measurements of arcs of the meridian in different latitudes were somewhat contradictory. The Cassini school asserted that the earth was elongated towards the poles. In order to decide this question, the French Academy sent an expedition under Bouguer to Peru in 1735 and another under Maupertuis to Lapland in 1736, to measure the length of a degree of the meridian. The results of these expeditions, combined with measurements of the length of a degree arc in France, proved that the length of a degree of the meridian increased as the latitude increased and established the correctness of Newton's view that the earth was flattened towards the poles.

The first observations of the positions of stars in the southern sky were made from St. Helena by Halley, who in 1678 published a catalogue of the places of 341 southern stars. Much more extensive observations were made by Lacaille in the course of a scientific expedition to the Cape of Good Hope, 1750–54, organized by the French Academy. More than 10,000 stars were observed with care. A catalogue of 400 of the brightest stars was published in 1757; a larger catalogue of 2000 stars was published posthumously in 1763. The positions derived from the observations of the remaining 8000 stars were first published in 1845 by the British Association. Lacaille also observed and described a number of southern nebulæ, nebulous stars, and star clusters.

In 1705 Halley published his 'Synopsis of Cometary Astronomy.' He had collected all available observations of 24 bright comets, whose parabolic orbits he computed. Struck by the resemblance between the paths of the comets of 1531, 1607, and 1682 and by the approximate equality in the intervals between their respective appearances, he conjectured that the three comets were different appearances of the same comet and that comets really moved in elongated elliptical orbits under the gravitational attraction of the Sun. The inequalities in the intervals of appearance he correctly attributed to the perturbations produced by planets near which the comet had passed; making approximate allowance for the perturbations produced by Jupiter he predicted that the comet would return about the end of 1758. The comet, known as Halley's comet, was first detected by an amateur astronomer, Palitzch of Saxony, on Christmas Day, 1758. Great interest had been aroused by Halley's prediction, as the predicted time for the reappearance of the comet drew near. The return of the comet proved that the motions of these bodies were in accordance with the law of gravitation and not only shook the superstitions attaching to comets but also convinced those who still hesitated to accept Newton's theories.

The Greeks had distinguished between the fixed stars, which were attached to a crystal globe, and the wandering stars or planets. After the Copernican system was accepted, it was realized that the stars were scattered through space, at various distances from the earth. There was no longer any reason to suppose that they were fixed. In 1718 Halley showed, by comparing the positions provided by current observations with those obtained in Greek times, that a few of the bright stars had changed their positions; the stars were not, in fact, fixed. The derivation of the proper-motions of the stars, from the comparison of positions separated by an interval of years, became of importance. The improved accuracy in the determination of the positions of stars from the time of Flamsteed onwards was gradually providing material suitable for this purpose. In particular, Maskelyne at Greenwich made numerous careful observations of the positions of 36 bright stars, for which he derived proper-motions. Herschel, using the derived proper-motions of 14 bright stars, showed in 1783 that the motions were not distributed at random; they showed a systematic effect which could be attributed to a motion of the Sun in a direction towards the constellation of Hercules. Though this conclusion was based on scanty data, its general correctness has since been fully substantiated.

Another discovery made by Halley was to prove of great theoretical interest. From the study of a number of ancient eclipses of the Sun and Moon he came to the conclusion that they could be reconciled with the motion of the Moon derived from current observations only by supposing that the mean motion of the Moon was being accelerated. More than two centuries were to elapse before the causes of this secular acceleration of the Moon's mean motion were fully understood.

The observations of the positions of the Sun and Moon which were carried on systematically at the Greenwich Observatory, combined with observations by Lacaille, Tobias Mayer, and others, were providing the data required for the construction of improved tables of the motions of these bodies, important not only for astronomy but also for navigation. Lacaille in 1758 published new solar tables, in the construction of which the perturbations by the planets were for the first time taken into account. These tables were revised and improved by Tobias Mayer (1723–62), director of the Göttingen Observatory, and the most important astronomer produced by Germany during this period. Mayer's most important work was on the Moon; he studied its librations and the theory of its motion, and succeeded in calculating lunar tables which were a considerable improvement on any earlier tables. In proceeding from the mathematical theory of the Moon's motion to the construction of tables, various numerical quantities must be derived from observation. The superiority of Mayer's tables was due in large measure to the skill with which he deduced these numerical quantities. Mayer sent his tables to England in 1755, where Bradley reported favourably on their accuracy. Revised tables, prepared by Mayer shortly before his death, were also sent to England. His 'Theory of the Moon' and his improved Solar and Lunar Tables were published in England in 1770 at the expense of the Board of Longitude.

The urgency of the solution of the problem of finding longitude at sea had become so great that in 1713 the British Government had offered a prize of £20,000 for a method that was reliable to within half of a degree, together with smaller prizes for methods of less accuracy. A Board of Longitude was appointed to consider applications for the prizes. The prize of £20,000 was eventually won by John Harrison for his invention in 1759 of a time-piece that would keep accurate time at sea, which was the forerunner of the modern marine chronometer. Before the end of the century, largely through the work of John Arnold (1736–1799) and of Thomas Earnshaw (1749–1829), the chronometer had been brought to a form which is substantially identical with that of the present day.

The invention of the marine chronometer did not do away with the necessity of determining longitudes at sea by the method of lunar distances. The astronomical observations were still needed in order to provide a general check on the gaining or losing rate of the chronometer. Maskelyne, who became Astronomer Royal in 1765, took a special interest in the problems of navigation. In 1763 he had published the 'British Mariner's Guide,' a handbook for the determination of longitude at sea by the method of lunars. The year after his appointment as Astronomer Royal, he published the first number of the 'Nautical Almanac' (for the year 1767), designed for the use of seamen, in which the astronomical data were given in the most convenient form. Maskelyne continued to produce the Almanac for 44 years and it justly acquired a high reputation for accuracy. Its publication has continued uninterruptedly to the present day but necessarily with numerous changes in form and content. Amongst other work carried out by Maskelyne, mention may be made of his measurement, in 1774, of the deflections of the plumb-line on each side of the mountain of Schehallion in Perthshire, which provided, through calculations carried out by Charles Hutton, the first determination of the mean density of the earth or, what is equivalent, of its mass. Near the end of the century, in 1798, an improved determination was made by Henry Cavendish by means of the well-known Cavendish experiment.

In June 1761 and 1769 occurred the rare phenomenon of a transit of Venus across the face of the Sun. Of the previous pair of transits, in December 1631 and 1639, the second had been observed by Jeremiah Horrox, the first such transit ever to be seen. The next pair were not due to occur until December 1874 and 1882. Halley had pointed out in 1679 that the transits of Venus could be utilized for determining the distance of the earth from the Sun by observing from different parts of the earth's surface the apparent path of the planet across the disk of the Sun. Extensive plans were made by governments, academies, and private persons for the observations of the transits. Expeditions were sent from many countries to various parts of the earth, from Siberia to California, from Hudson's Bay to the Cape of Good Hope and Madras.

from Norway to Tahiti. The occasions provided the first instance of international co-operation on a large scale in astronomy. Both transits were widely observed, but the results proved disappointing. It had been expected that the internal contacts of Venus with the Sun could be timed with high accuracy. This proved not to be the case ; disturbing effects were caused by the atmosphere of Venus, so that the times of the contacts determined by different observers were singularly discordant. Widely different values for the solar parallax were obtained by discussing the observations in different ways, indicating the presence of systematic errors. No real improvement in the knowledge of the distance of the Sun resulted from the immense trouble taken over the observation of these two transits.

The eighteenth century saw considerable developments in the design and construction of telescopes. In 1720 Hadley made the first really workable reflecting telescope. It had a 6-inch speculum mirror, of 63-inches focal length, mounted in a wooden tube and capable of motion in altitude and azimuth. Hadley cast, ground, figured and polished the mirror himself, taking successfully the essential step of parabolizing the mirror, which had never been achieved before ; he also developed methods for testing the figure as the work proceeded. Bradley tested Hadley's telescope against a Huygens objective of 123 feet focal length and found that it would bear as much power with equal definition. Hadley disclosed his methods of working to others ; Molyneux, Hawksbee, and particularly James Short soon acquired a high reputation for the excellent qualities of the reflecting telescopes which they made.

Attention had been turned to the reflecting telescope because of the optical aberrations of simple lenses, which compelled long focal lengths to be used, with all the resulting inconveniences. In 1733 the achromatic lens was invented by Chester Moor Hall, an Essex magistrate, who had some made for him by a London optician, George Bass. In 1758 it was independently invented by John Dollond, who took out a patent for it. Dollond was an excellent craftsman and his small achromatic telescopes of about 5 feet focal length had a ready sale.

Improvements were also made in the mountings of telescopes. The compact size of the telescopes with achromatic objectives made the equatorial form of mounting practicable. Sisson in 1760 devised the simple English mounting, with the polar axis supported by an upper and a lower bearing. Before the end of the century Ramsden had introduced the English yoke mounting, in which the polar axis is in the form of a yoke or cradle, the two sides of which support the declination axis and within which the telescope swings. The Shuckburgh equatorial, made by Ramsden in 1791, was a marked advance on earlier construction.

The achromatic objectives were limited in size, however, by the poor quality of the glass which was available. It was far from homogeneous, the flint particularly being full of faults. Pieces of glass suitable for lenses of apertures greater than 3-inches were rarely obtained. For greater light-gathering power, speculum mirrors still had to be used. William Herschel (1738–1822) was the first to make telescopes of large aperture. In 1774, after he had settled in Bath, he undertook the construction of his first reflector, which was of the Gregorian type. He had to learn by hard experience the difficult technique of casting, figuring, and polishing the mirrors. The figuring of the Gregorian secondary mirror proved difficult, so he soon gave preference to the Newtonian type. As his skill and technique improved, he was able gradually to increase the size of his mirrors, adhering usually to the rather large f-ratio, for a reflector, of about 12 or 13. The best of his mirrors were of extremely good quality; one of $6\frac{1}{4}$-inches aperture, of 7 feet focal length, was tested at Greenwich against a $9\frac{1}{2}$-inch mirror of Short's and found to be superior to it. On 13 March, 1781, when examining the heavens with his best 7-foot telescope, he detected an object which showed a minute disk. He thought at first that it was a comet, but it proved to be a new planet, the first planet to be "discovered," to which he gave the name of the *Georgium sidus*. This name was not in accordance with the practice of naming the planets by the deities of classical mythology, and the name Uranus, which had been suggested by Bode, was soon adopted in preference to the name Herschel had assigned.

The discovery of the new planet enlarged the bounds of the solar system and brought Herschel fame. He was appointed King's Astronomer by George III and moved to Slough, where he was able to devote himself entirely to astronomical observation and to the making of telescopes. After several unsuccessful attempts to cast larger mirrors, he succeeded by the autumn of 1783 in completing two good mirrors of $18\frac{3}{4}$-inches diameter and 20-feet focal length. The 20-foot telescope was his favourite instrument, with which most of his subsequent observations were made. It was too large to be mounted at that date as an equatorial, so that the less convenient altazimuth form of mounting was used; the telescope was mounted between the two parts of a light framework, supported from a circular turntable. It could be raised or lowered in altitude by tackle, the turntable being rotated to provide motion in azimuth.

Herschel made and sold a large number of telescopes, up to the 20-foot size, to amateurs in all parts of Europe, to kings, princes, and financiers; the receipts enabled him to carry out costly experiments on the mechanical grinding and polishing of large mirrors. He designed a large polishing machine, which was finished in 1788. Several unsuccessful attempts were made to cast a speculum disk larger than any yet made. At length success was achieved and a large telescope, of 48-inches clear aperture and 40-feet length, was constructed. Its aperture-ratio was the largest of any of the telescopes made by Herschel. An altazimuth mounting, which was a much elaborated form of that used for the 20-foot telescope, was employed. The telescope was not of the Newtonian type but was arranged as a front view telescope, the mirror being slightly tilted, in order to throw the image near to the top outer edge of the tube. The large f-number made it possible to tilt the mirror sufficiently without introducing too much coma, while the loss of light caused by reflection at a secondary mirror was avoided. With this telescope Herschel discovered the 6th satellite of Saturn, Enceladus, and also the 7th, Mimas. This great telescope proved to be of value in sweeping for nebulæ and for studying faint objects. But it was used relatively little, for it was rather cumbrous to manipulate; there was some trouble from flexure of the mirror under its great weight, the method of supporting it being rather primitive; also the rapid tarnishing of the mirror necessitated repolishing, which involved refiguring, every two years. In order to facilitate the casting, annealing and working of the large speculum disks, Herschel increased the copper content to 75 per cent., with the result that the surface tarnished easily.

Another instrumental development of importance was the invention of the heliometer by Dollond in 1755. In this instrument the objective is divided into two halves along a diameter, the two halves being capable of separation along the common diameter, the amount of separation being indicated by a scale. The objective and mount could be rotated into any position angle. The distance between two stars is determined by bringing the image of one star, formed by one half of the objective, into coincidence with the image of the other star, formed by the other half of the objective, when the common diameter has been set along the direction of the line joining the stars. Distances up to about 2° can be measured with accuracy, but the instrument was designed primarily, as its name implies, for the measurement of the Sun's diameter. Its principal application at a later date was for the determination of stellar parallaxes; the accuracy of the results obtained with it was not surpassed until the development of photographic methods caused the heliometer to be superseded.

Some outline must now be given of the more important developments in gravitational astronomy during the eighteenth century. The general problem of determining the motions and positions of three or more bodies under their mutual gravitational attractions is insoluble, except for a few special cases. In the most important cases which arise in the solar system there are special circumstances which enable solutions to be obtained by successive approximations. In planetary theory, the masses of the planets are very small compared with the mass of the Sun, so that the forces due to the other planets which perturb any particular planet are relatively small; also the eccentricities and inclinations of the planetary orbits are all small. In the more difficult lunar theory, the great distance of the Sun relative to that of the earth provides a simplification. Much attention was given to the theory of the Moon's motion because of its importance for the determination of longitude at sea. Euler, Clairaut, and d'Alembert occupied themselves with the problem and derived approximate solutions independently and nearly simultaneously. They all deduced from their respective analytical solutions a motion of the apogee that was equal to only half the observed motion. As Newton had encountered the same discordance between theory and observation, it began to be suspected that gravitation did not vary exactly as the inverse square of the distance. In 1749, however, Clairaut found that when certain small terms, which had previously been neglected, were taken into account, the computed and observed values were brought into agreement. Clairaut's lunar theory was published in 1752, Euler's in 1753, and d'Alembert's in 1754. The employment of analytical methods enabled the principal inequalities in the motion of the Moon, which were known from observation, to be explained with considerable accuracy quantitatively as well as qualitatively, with the one important exception of the secular acceleration of the mean motion, which had been discovered by Halley. Clairaut's most important work was his treatise on the figure of the earth, published in 1743. In this work he gave for the first time the general equations of the equilibrium of a fluid and applied them to the determination of the form which a rotating body like the earth would assume under the influence of the mutual gravitation of its parts, hypotheses of a very general nature as to the variations of density in its interior being made. He deduced a formula for the value of gravity in different latitudes which is in close agreement with the results of pendulum observations. Clairaut's treatise has formed the basis for all subsequent investigations of this subject, to which few results of fundamental importance have since been added.

The first attempts were also made about this time to account for the inequalities in the motions of the planets caused by perturbations by other bodies. The most important were the inequalities in the motion of the earth, which showed as irregularities in the apparent motion of the Sun; and certain irregularities in the motions of Jupiter and Saturn, which had been known to Horrox and which Halley had suggested were caused by their mutual perturbations. Clairaut, in 1757, discussed the perturbations of the earth by the Moon, Venus and Jupiter; by comparison of his theory with observation, he was able to estimate the masses of Venus and the Moon. His mass of the Moon was a considerable improvement on Newton's estimate from the tides, while the mass of Venus was not only new, but was in fair agreement with modern estimates. This was the first attempt to estimate masses of celestial bodies by means of the perturbations which they produced.

Euler made an important contribution to general theory by his method of the variation of the elements or parameters of an orbit, which, though developed in 1756, was not published until 1771. A planet moving in a perturbed orbit was represented as moving at any instant in an ellipse, whose elements are continually changing. The whole effect of the perturbations was in this way thrown upon the elements of the orbit; when these elements had been determined for any instant of time the position of the planet could be calculated by ordinary elliptic theory.

The important contributions made to gravitational astronomy by Lagrange and Laplace, the two greatest mathematicians of the next generation, commenced when the work of Clairaut, d'Alembert, and Euler was nearly finished. There was an important difference between these two great contemporaries. Lagrange valued elegance and generality of treatment ; a particular problem was an occasion for the application of a general method. Laplace used mathematics as a tool, to be modified to suit each special problem. Lagrange made the more important contributions to general dynamical theory, whereas Laplace made the more important contributions to celestial mechanics in so far as the detailed explanation of the motions of the celestial bodies is concerned. They both assisted in the development of the mathematical treatment of perturbations, distinguishing between those which are *periodic* and those which are *secular*. When the perturbations of one planet by another are investigated, the mathematical expression for the disturbing force depends upon the positions of the planets at a given time and also upon the elements of their two orbits. It can be divided into two parts, one of which involves the positions of the planets while the other does not, though both involve the orbital elements. The first portion of the disturbing force changes relatively rapidly, going through a complete cycle of changes in an interval of time which is of the same order as the orbital periods of the planets. This part can be represented by a sum of periodic terms. The second part changes very slowly, as the elements of the orbits change : for long periods of time, this part can be represented by a power series, with the time as the variable quantity. It is generally sufficient to use only the first three terms, which represent the value at a particular epoch, the rate of change at that epoch, and the secular change or acceleration. In lunar theory the periodic inequalities are large and of great importance ; in planetary theory the periodic inequalities are generally small and the secular inequalities are important.

The question arises whether the mutual perturbations of the bodies in the solar system could be cumulative to such an extent that the system would become unstable. Lagrange and Laplace both contributed to the solution of this problem. Laplace in 1773 proved that the axes of the planetary orbits were not undergoing secular change, nor could they have altered appreciably for a long time past ; Lagrange in 1774 proved that the inclinations of the orbits could not pass outside certain limits. Then in 1775 Laplace obtained a similar result for the eccentricity. A further step forward was made by Lagrange in 1776, who proved that, provided the disturbing forces were small, the lengths of the axes could only fluctuate within certain limits, whatever the eccentricities, inclinations and masses might be. The final step was made in 1784 by Laplace, who proved that, whatever the relative masses of the planets, the eccentricities and inclinations, if once inconsiderable, would always remain so provided that the planets all revolved round the Sun in the same direction. The conclusion from this series of investigations is that the changes in lengths of axes, in eccentricity, and in inclination of any planetary orbit are all permanently restricted within certain definite limits. The solar system is stable.

The slow alteration in the rates of motion of Jupiter and Saturn, to which Halley had drawn attention, proved to be difficult to explain. Lagrange in 1766 attempted unsuccessfully to account for it as a secular inequality. Laplace, in 1773, by a more complete analysis showed that there was no secular change in the motions of these planets. He eventually found the real source of the irregularities in the near commensurability of the mean motions of the two planets, which are approximately in the ratio of 5 to 2. In consequence certain terms in the longitudes which would normally be small, being of the order of the cubes of the eccentricities, become larger because they are divided by the small factor $(5n-2n')^2$, where n, n' are the mean motions. An appreciable inequality, with a period of 929 years, is thus produced. The observations were satisfactorily accounted for and the irregularities of the two planets, which had seemed to be inexplicable by the law of gravitation, provided a striking proof of its truth.

Laplace succeeded also in finding the cause of the acceleration of the mean motion after both Euler and Lagrange had failed in their attempts to account for it. As a result of the action of the planets on the earth, the eccentricity of the earth's orbit is decreasing slowly. In consequence there is a gradual decrease in the mean action of the Sun upon the Moon, which causes a slow increase in the Moon's mean motion. Laplace calculated the rate of increase and obtained a value in close agreement with observation. His memoir on this subject was published in 1788 and appeared finally to have settled the question. But, in 1853, J. C. Adams showed that the calculations of Laplace were incomplete, and that the correct theoretical value was little more than half the observed value. Some seventy years were to elapse before G. I. Taylor and H. Jeffreys proved that the residual effect was attributable to the slowing down of the earth's rotation through tidal friction.

In his great work, the 'Mécanique Céleste,' published in five volumes over a period of 26 years (1799–1825), Laplace presented all his astronomical work as a reasoned whole, and exhibited the theory of the solar system in a state of almost complete development. This work was highly mathematical; to the layman Laplace is best known through his nebular hypothesis of the origin of the solar system, which was published in 1796 in his popular book, the 'Système du Mond.' Laplace put the hypothesis forward merely as a conjecture with, as he said, "all the distrust which everything which is not a result of observation or of calculation ought to inspire."

We have seen that great progress was made during the eighteenth century in positional astronomy and in gravitational astronomy. Equally great progress was made in descriptive astronomy, the branch of astronomical observation which is concerned with questions of the appearances and structure of the celestial bodies and of their mutual relations. We are here concerned almost entirely with the work of one man, William Herschel, whose most important researches fall into the last quarter of the century. Mention has already been made of his skill in the construction of telescopes and of his discovery of the planet Uranus. After settling at Slough, Herschel was able to devote himself assiduously to observation and for many years he observed throughout every suitable night. In 1811 he said that " a knowledge of the construction of the heavens has always been the ultimate object of my observations." Beyond the limits of the solar system, astronomers had hitherto concerned themselves only with the measurement of the positions and motions of the stars ; several attempts had been made to measure the distances of stars, but with no success. They had never studied the stellar universe *per se*, so that Herschel in these observations was opening a new chapter in astronomy.

Because the attempts of Bradley and others to measure stellar distances had failed, Herschel decided to attack the question of stellar distribution by an indirect method. For this purpose he developed his method of star-gauging, counting the number of stars visible in the field of his telescope (a circular area 15′ in diameter) in different regions of the sky. The variation in the numbers of stars seen might be due either to a real inequality of the distribution of the stars in space, or to the stellar system extending to greater distances in some directions than in others, or to both causes combined. Herschel assumed that the distribution of the stars was approximately uniform and was then able to estimate the relative extension of the system in different directions. The true nature of the Milky Way had been a matter of conjecture before he commenced his observations. He proved that it was an aggregation of a vast number of faint and therefore distant stars, which were individually too faint to be visible to the naked eye. He found that the space occupied by our stellar system was shaped roughly like a disk or grindstone, the plane of the disk corresponding with the central plane of the Milky Way. The diameter of the system was about five times its thickness and the Sun was somewhere near its centre.

Herschel was aware that the hypothesis of uniform distribution was at best merely an approximation. The existence of star clusters showed that it was not correct in detail. He therefore made use also of the brightness of the stars as a measure of their relative distances. If the stars have the same intrinsic brightness, a knowledge of their apparent brightness would enable their relative distances to be inferred. At that time stars were assigned a magnitude merely by comparison with adjacent stars to which magnitudes had already been assigned. Photometric observations to compare the relative brightness of stars of different magnitudes had not yet been made. Herschel made rough comparisons of the amount of light received from different stars by using telescopes of different apertures and also by measuring the aperture at which a star just remained visible.

In one of his latest papers Herschel (1817) made an attempt to compare the assumptions of uniform distribution and of uniform intrinsic brightness, and concluded that the fainter stars are either more closely packed in space than the brighter or that they are intrinsically less luminous.

In the course of his systematic surveys of the heavens, Herschel made a note of everything that was unusual. In this way he recorded a great number of nebulæ and star clusters. Messier had recorded a certain number of nebulæ in the northen sky and in 1781 published a list of 103 objects, his purpose being to prevent such objects being mistaken for comets by those who were engaged in comet searching. Lacaille had published a list for the southern sky. In 1786 Herschel published a list of 1000 new nebulæ and clusters; in 1789 a further list of 1000 objects; and in 1802 a third list containing 500. The appearance and position of each object was described.

Herschel gave much thought to the nature of the nebulæ. He was for some time inclined to the belief that they were aggregations of faint distant stars which his telescopes could not resolve into discrete stars, just as the naked eye cannot resolve the Milky Way into discrete stars. But the detection in 1791 of a star surrounded by a circular nebulous halo convinced him that some of the nebulæ were not aggregations of stars, but consisted of " a shining fluid, of a nature totally unknown to us." The possibility that other nebulæ might be distant clusters of stars was not excluded. He noted that objects of different types were differently distributed over the sky; nebulæ associated with clusters were common near the Milky Way and were often surrounded by a region of the sky relatively devoid of stars. (These are now known to be gaseous nebulæ, shining by the light of stars embedded in them, and associated with clouds of obscuring dust.) The other type of nebulæ, devoid of stars, were most frequent near the poles of the galactic circle. (These are the extra-galactic nebulæ, whose density decreases towards the Milky Way because of galactic absorption.) Up to the end of his life Herschel adhered to the view that some of the nebulæ were external universes, but that others belonged to our galactic system. The correctness of this view was not established until 1920.

The discovery of binary stars, or physical double stars, was made by Herschel. He had prepared a catalogue of pairs of stars lying nearly in the same line of sight, with angular separation not exceeding 2′. His intention was to keep these stars under observation in order to determine their relative parallax. Supposing the proximity of the stars to be merely accidental, it is probable that the fainter star of the pair is the more distant, so that the parallactic displacement, caused by the revolution of the earth round the Sun, should be less for the fainter star than for the brighter star. Herschel therefore measured the distance between the stars and the position angle of the line joining them. On repeating his observations at later dates, he found for a few of the double stars changes in distance and position angle which were not such as could be explained by a relative proper-motion, and could only be due to the two stars revolving round one another under the influence of their mutual gravitation. These stars must consequently be true binary systems. The binary nature could be conclusively established in Herschel's time for a few only of the double stars, though changes of position angle and distance were recorded also for others. No success in determining stellar parallaxes was achieved; we now know that the parallactic displacements were too small to be detected. But a more important discovery was made, for it is only through the gravitational interaction of stars in binary systems that any information about stellar masses can be obtained.

The discovery of the binary nature of some of the close pairs of stars verified the conclusion which had been reached by John Michell from considerations of probability. Considering the distribution of stars over the sky Michell showed in 1767 that the number of close pairs was far larger than was to be expected from a random distribution. In 1784, after some of Herschel's lists of double stars had been published, Michell returned to the question and concluded that many of the pairs must be physically connected.

Because of the difficulty of making accurate photometric measurements, Herschel devised the simple method of arranging groups of stars in sequences in order of relative brightness as a means of detecting at a future time any marked amount of variation of brightness. He used this method of sequences extensively and between 1796 and 1799 published four catalogues of comparative brightness, detecting in the course of the work a number of variable stars.

Besides the stellar observations, Herschel added considerably to the knowledge of the solar system. In addition to discovering Uranus and two satellites of Saturn, as already mentioned, he discovered two satellites of Uranus, showed that their orbits are nearly perpendicular to the ecliptic and that their motion is retrograde; he determined the period of rotation of Saturn, discovered its spheroidal figure, and measured its ellipticity; he detected the rotation of the ring system, discovered the faint inner crape ring, and showed that the thickness of the rings did not exceed a few hundred miles; he found that the variations in brightness of Iapetus were periodic, the period being equal to the period of revolution of the satellite; he also noted the seasonal changes in the polar caps of Mars.

The researches of Herschel opened up so many new lines of investigation that he has justly been termed "the father of modern astronomy." When the nineteenth century began, there were already living several who were destined to make important contributions to astronomy—Gauss, Bessel, F. G. W. Struve, Argelander, Hansen, Fraunhofer, and John Herschel. Gauss was already developing his method of determining the orbit of a planet or comet from three observations. The first of the minor planets, Ceres, was discovered on January 1st, 1801, only to be lost after having been under observation for a few weeks. Gauss applied his new method to determine its orbit, with the result that the planet was found again at the end of the year. His method of least squares was to prove of the greatest value in the analysis and adjustment of astronomical observations. Bessel was to systematize the reduction of meridian observations, to construct much improved tables of refraction, and to set a new standard for precision of observation. The accurate observations of double stars continued by Struve for many years at Dorpat and Poulkova were to establish the foundations of double-star astronomy on a secure basis. Bessel, Struve, and Henderson were each to determine the distance of a star and, practically simultaneously, to succeed where so many of their predecessors had failed. Argelander was to prepare the great catalogue of some 324,000 stars and the star maps of the ' Bonner Durchmusterung,' a work of the greatest reference value to astronomers. Hansen was to make a great advance in lunar theory and to construct tables of the Moon that were a great improvement on all previous tables. Fraunhofer, by his careful examination of the dark lines in the Sun's spectrum, was to show that new knowledge was to be gained by the application of the spectroscope to astronomical observation, thereby paving the way for the birth of a new branch of astronomy, called astrophysics, which is concerned with the physical conditions, the composition and the constitution of the celestial bodies. His collaboration with the Swiss artisan, Guinand, was to lead to important developments in the manufacture of optical glass. Guinand had commenced in 1784 to improve the quality of optical glass, and by 1799 had succeeded in making flint disks of high quality, as large as 6 inches in diameter, and also in producing denser types of flint. For some years from 1805 he joined forces with Fraunhofer : Guinand produced the glass, Fraunhofer devised the methods of working it on automatic machines. These developments were to make it possible for achromatic objectives of larger size to be constructed. Fraunhofer was, in addition, to perfect the equatorial mounting and its clockwork drive, and to bring the refracting telescope substantially into its present form, the reflecting telescope being almost entirely superseded for about half a century. John Herschel was to extend his father's survey of the heavens to the southern hemisphere and to add greatly to the knowledge of star clusters, nebulæ, and double stars.

Besides the great names we have mentioned there were many others who were born before the nineteenth century opened and whose work was to have important significance. Among them were Schwabe, who was to discover the periodicity of sunspots, and Sabine and Gautier, who were to show that there was a relationship between the frequency of the appearance of sunspots and various magnetic phenomena on the earth. Lord Rosse, who in 1845 constructed and erected at Birr Castle the great reflector of 6-foot aperture and 54-feet focal length, was born in 1800. This great instrument, with its cumbrous method of mounting, belongs in a sense to the eighteenth century, for the mirror was the last large speculum mirror to be used. But it has to its credit one important and significant discovery, that of the spiral structure of many of the nebulæ which are now known to lie beyond our galaxy, and thereby it forms a link between the observations and speculations of William Herschel in the eighteenth century and the new knowledge that has been gained about these objects in the twentieth century.

One of the most sensational triumphs of astronomy in the nineteenth century was the discovery in 1846 of the planet Neptune in the position that had been indicated by Leverrier and Adams. Both of these astronomers belong to the nineteenth century; but it was because the work of Lagrange and Laplace during the eighteenth century had placed the law of gravitation on so firm a basis that it became reasonably certain that the discordances between the observed and computed positions of Uranus (itself a discovery of the eighteenth century) must be due to the perturbing action of an unknown planet, and both Adams and Leverrier were thereby induced to attack the difficult problem of inverse perturbations.

PHYSICS IN THE EIGHTEENTH CENTURY

By Prof. HERBERT DINGLE, D.Sc.

At the time when the 'Philosophical Magazine' was founded, a reflective natural philosopher, looking back on the century which was closing, would have beheld a scene of some complexity. The achievement of the seventeenth century had been precise profound and almost immeasurably great. At the beginning of that great epoch the conception of the universe which, though not unchallenged, was still predominant, was essentially mediæval in spirit ; at the centre were the earth and the baser elements, subject to change and decay and tending to move in straight lines towards their own places, while surrounding them the crystalline spheres, incorruptible in the heavens, performed eternally the perfect circular motions proper to their nature. By the end of the century the whole scheme had vanished and the Newtonian system reigned in its stead. The spheres had disappeared ; matter and motion were the same everywhere, on the earth and in the heavens, and were indissolubly linked with one another through the universal force of gravitation ; and a technique had been created for the conquest of other phenomena by means of Newtonian forces yet to be discovered. After a change so radical it was only to be expected that a period of development would set in, a period on one hand of full exploitation of the field of mechanical motion for which Newton had provided the means, and on the other of the application of his essential principles to other physical phenomena such as heat, electricity, magnetism and the rest which he had left practically untouched. The former effort belongs mainly to astronomy, which provides almost completely perfect examples of the operation of Newtonian forces ; the latter lies chiefly in the realm of physics and is the subject of our present concern. It lasted, in fact, for two centuries, and midway in that long period the 'Philosophical Magazine came to birth. Of the achievements of the nineteenth century it is itself no bad record. Let us place ourselves at the beginning of that time and survey the heritage which the eighteenth century had bequeathed to it.

I have said that the view which faces us would have appeared to a natural philosopher of the time to be a complex one. That is true, and if today we are able to disentangle its threads and see a simple pattern running through it, the reason is that we have now left the Newtonian epoch sufficiently far behind to understand in large measure what its true significance was. We see it in a light which did not shine for those who worked therein, and this fact, fortunate though it is for our understanding of science, has compensating disadvantages when we try to understand scientists. The power to clear one's mind of knowledge and of tendencies that have become almost instinctive and to live the mental life of a dweller in days long past is not a possession of the present writer, and he will not make the vain effort to acquire it. This survey of eighteenth century physics is a distant retrospect, an interpretation as well as a description, and it makes no pretence to be anything else.

It should not be without profit, however, to begin by reminding ourselves of a peculiar difficulty which besets our understanding of the early post-Newtonian natural philosophers and makes them in one sense more alien from us than the men of the Middle Ages. Mediæval science, strange and indelibly coloured with enchantment as it appears to us now, was at least a unity. Modern divisions, such as those separating astronomy from theology, and both from psychology, did not exist. The Abode of the Blessed was outside the remotest stellar sphere, and the planets influenced men's temperaments and terrestrial fortunes while at the same time declaring the glory of God. Such progress as was made was all within the framework of one unifying idea, and if one succeeds in becoming mediæval in one sphere of thought he is not far from entering completely into the spirit of the age. But with Galileo and Newton this unity disappeared for ever. The history of thought, which could formerly proceed along a single line though at an irregular pace, becomes broken up into separate courses, like a river entering its delta. One branch—the study of mechanical motion—far outstrips the others, which themselves advance at various speeds. To understand the thought of a particular time we must take a cross-section, and this reveals the most incongruous associations. At the end of the eighteenth century, for instance, we find Sir William Herschel entertaining views of the stellar system and the distant nebulæ very similar to our own, and at the same time holding that the Sun was an inhabited globe whose people were protected by a cloudy veil from the hot outer shell ; while in France, Laplace was calculating how the curvature of space was related to the matter in it and the Academy was refusing to consider evidence for the fall of meteorites on the ground that such events could not possibly happen. Who will undertake to recapture the outlook in which these are co-existing elements ?

The dominant fact that directed the whole of eighteenth century science was that Newton had succeeded in formulating laws of motion so complete and comprehensive as to be applicable to phenomena large and small throughout the universe. This having been achieved, not only dynamical astronomy, but also the whole field of terrestrial mechanics lay at the mercy of the natural philosopher, who had need of mathematical, rather than experimental, skill to reap the richest harvests. His success was great and lasting, and is on this account better represented in modern text-books than the now largely superseded pioneer work of the students of other departments of physics. For this reason it would be an unprofitable use of a limited space to describe the contributions of the eighteenth century to the study of the general mechanical properties of matter and of sound, and I shall, therefore, confine myself to those subjects in which the

Galilean-Newtonian philosophy stood as an example to be copied rather than a fully efficient tool to be directly applied. Its essence lay in its isolation of a phenomenon—motion—exhibited by bodies of all kinds, in whatever other characteristics such as shape, colour, temperature and what not they might differ, and the choice of concepts—space, time, mass and force—in terms of which the laws of motion could be described. If, then, other phenomena were to be dealt with in the same way, the first step was clearly the choice of the appropriate concepts, and this was the problem whose solution, though not then quite completely reached, was pre-eminently the work of the eighteenth century. The formulation of general laws in terms of those concepts was left to the nineteenth century.

Heat.

Let us begin with heat. Galileo had already conjectured that since, in his view, external bodies possessed no qualities other than size, figure, number and motion, the physical origin of our sensation of heat was " a multitude of minute corpuscles thus and thus figured, moved with such and such a velocity." But he made no experiments and hazarded no speculations concerning the figure and velocity of such corpuscles, nor did he make even the preliminary measurements necessary to suggest the kind of concepts which would have been appropriate to their description. He did, indeed, devise a "thermometer"—the first such instrument of which there is any record—but it appears to have been used only for purely practical ends. In his day such ideas as one can find concerning heat are of the vaguest.

We can perhaps realise the state of affairs by imagining ourselves as looking at our present clear-cut trilogy of thermal concepts—heat, temperature and entropy—through a sort of mental telescope which we can gradually put out of focus. The first effect is the merging together of heat energy and entropy into *caloric*, the puzzling character of which was largely due to the unrealised incompatibility between the conservation of the one and the continual growth of the other—a contradiction from which Carnot was saved by his choice of conditions of reversibility in which entropy also is conserved. A further turn of the screw takes us back through the eighteenth century and shows us a single blur in which caloric and temperature are indistinguishable. Nevertheless, to continue the metaphor, it is not a uniform blur. Two suggestions of foci are discernible, interpreted as opposite qualities of " heat " and " cold." Such was the view of Morin, a younger contemporary of Galileo, who, apparently uninfluenced by Galileo's inspired but unfruitful conjecture, regarded heat and cold as separate but inseparable entities. What we would now call a temperature of x degrees was to Morin a mixture of a degrees of heat and b degrees of cold, where $a+b=8$ (apparently a conventional figure resulting from the custom of dividing the range of the thermometers of the time into eight parts) ; and with this picture in his mind he discusses the effect of mixing together quantities of water at different temperatures. For this purpose he adopts certain principles (apparently purely hypothetical) governing the interaction of the opposite qualities, and obtains, in fact, surprisingly good results *.

It is fortunate that Newton's example of empiricism here supervened to stimulate accuracy of measurement and that, without waiting to know exactly what they were measuring, physicists turned their attention to the improvement of the "thermometer." By the end of the seventeenth century, liquids, such as water or alcohol, had taken the place of Galileo's air as the expanding substance, and the use of two "fixed points" and a standardized "degree" had been suggested though not practised. It was Fahrenheit who constructed the first thermometer worthy to be counted as a measuring instrument in the modern sense of the term. By 1721 he had substituted mercury for alcohol and had established the principle of the scale of what we now call temperature which still bears his name. His work was apparently suggested by a visit in 1708 to Römer, who had constructed a thermometer according to which the temperatures of melting ice and normal human blood were numbered respectively $7\frac{1}{2}$ and $22\frac{1}{2}$. Fahrenheit adopted this scale, with each division subdivided into four parts so that the afore-named temperatures became 30 and 90 degrees, respectively, and adhered to it until 1717, when, finding the fractions awkward, he increased the readings by one-fifteenth, so that the temperatures of melting ice and blood became 32 and 96. There appears to be no foundation for the common statement that Fahrenheit's zero defined the temperature of a mixture of ice and salt. By means of his thermometer he was able to verify the constancy of the readings corresponding to the melting of ice and the boiling of water (212), but he never regarded these as fixed points, and succeeded, indeed, by the use of his instrument, in showing that the boiling point of water varies with the atmospheric pressure.

Simultaneously, but independently, Réaumur in France was devising an instrument of comparable precision. Preferring alcohol to mercury because of its greater expansibility, he differed from Fahrenheit also in a much more fundamental, but practically less important, respect by fixing only one point (the zero, corresponding to the freezing point of water) and choosing as the degree the "heat" which made the alcohol expand by a definite fraction (generally $\frac{1}{1000}$) of its volume at the zero point. This entailed the laborious process of graduating the tube in terms of its volume, and introduced the principle, as we should express it nowadays, of making the dimensions of temperature those of volume instead of primary, or fundamental. This is of some interest in view of de Luc's improvement of the instrument later in the century, an account of which he published in 1772. By that time the idea of heat as a substance had become much more clearly established, and it was possible to determine

* Morin's statement is reproduced in 'The Discovery of Specific and Latent Heats,' by D. McKie and N. H. de V. Heathcote, to which I am much indebted.

whether Réaumur's degrees "really" corresponded to equal increments of "heat." Assuming that when two equal quantities of water at different "temperatures" (*i.e.* containing different amounts of "heat") are mixed, the average of the initial readings should be recorded by a "true" thermometer, de Luc showed that none of the thermometric liquids he tested reached this ideal when the graduations were made on Réaumur's principle, but that mercury came nearest to it. Réaumur, as we have seen, needed no upper fixed point, but he did, in fact, stipulate, as a standard of purity of the alcohol to be used, that it should expand from 1000 to 1080 units of volume when heated from the zero to its boiling point. This would, of course, indicate 80 degrees as the boiling point of alcohol, but in course of time this figure became attached to the boiling point of *water*. This matter also was cleared up by de Luc, and thenceforward, as now, the direct relation of expansion to volume was abandoned and the Réaumur degree was obtained by dividing the thermometer stem between the points corresponding to freezing and boiling water into 80 equal parts.

The third of the familiar standard liquid thermometer scales—that of Celsius—dates from 1742. There were two fixed points—the "temperatures" of melting ice and boiling water—and the interval was divided into 100 equal parts on the thermometer stem, but in the original instrument the higher point was zero and the lower was "100." Christin in the following year reversed this order and introduced our present "centigrade" scale.

So much for measuring instruments; but what did they measure? This was not a primary concern in the eighteenth century, and men like Black, for instance, were content to discover relations between their measures without asking themselves what was the nature of the things they measured. This mental discipline, learnt from Newton, is one which we are trying to restore in these days, after the Victorian striving for "models of nature," and we are less inclined than our grandfathers to think patronisingly of the earlier students of heat because they placed more emphasis on devising an accurate formula for calculating the thermometer readings when substances of different temperature were mixed than on the description of underlying physical processes.

Nevertheless, some mental picture was necessary even for this, and the purest empiricist could not avoid the choice of words to describe what was happening. To Morin, as we have seen, heat and cold were "qualities" measurable by their degree and not their amount, and the thermometer measured a sort of resultant quality like the "lukewarmness" with which we regard something that at the same time attracts and repels us. The eighteenth century, however, thought more in terms of one entity, variously conceived as a substance or a form of motion. To those holding the former view, the thermometer naturally measured the amount of heat in a body, but at the same time there was an inarticulate feeling that heat, whatever it might be, had both intensive and extensive aspects. Thus Richmann in 1747–48, in a memoir dealing with what we should call

" temperature " alone but which he entitled " On the Quantity of Heat " (De Quantitate Caloris), assumes that when the same " degree of heat " (idem caloris gradus) is uniformly distributed through different quantities of water, the " heat produced " (calor hinc generatus) is inversely proportional to their amounts. Two distinct ideas represented by the same word are here clearly evident.

Much the same confusion is found in Boerhaave, but in Black, who worked in the period 1759–63, we find the clearest possible distinction between the two ideas. He points out that Boerhaave fails to distinguish between quantity and intensity of heat and adds " very soon after I began to think on this subject (anno 1760) I perceived that . . . the quantities of heat which different kinds of matter must receive, to reduce them to an equilibrium with one another, or to raise their temperature by an equal number of degrees, are not in proportion to the quantity of matter in each, but in proportions widely different from this, and for which no general principle or reason can yet be assigned." Here clearly what the thermometer records is "temperature," and "quantity of heat" is something other. From this, Black goes on to define " capacity for heat " and " latent heat "—the name as well as the discovery of the fact is his. He concerned himself as little as possible with the problem of what " heat " might be, but inclined more towards the " substance " than the " motion " theory, as indeed the term " capacity " indicates. In his later years he tended to favour the theory advanced by Cleghorn in 1779, according to which the particles of heat, or " fire," repel one another but are attracted by different kinds of matter in different degrees. This accorded well with his experiments. Wilcke, who independently but later and less satisfactorily repeated Black's discoveries, also favoured the " substance " theory and conceived clearly enough that the " quantity of heat " in a body was something different from the reading of the thermometer when placed in contact with it. As to terminology, "fire," " heat," " matter of heat " and such terms were in use up to 1787, when the name " caloric " was coined by certain French chemists. This is the name by which the material theory of heat is generally known today, but it is considerably younger than the theory itself.

In so far as the pioneers of calorimetry were guided by theoretical conceptions, therefore, it is the material theory of heat that served their purpose. But there was another range of phenomena which forced itself on the attention of physicists in the latter half of the century, which showed at least that the question was not a simple one. Several experimenters demonstrated beyond question that when a red-hot body was placed at a focus of an optical system (generally a system of concave mirrors), a thermometer placed at the other focus showed a rise of temperature. Hence, something exhibiting the reflecting powers of light could appear as heat. It was not ordinary light because a plate of glass placed in the presumed course of the heating agency allowed light to pass but destroyed the heating effect, though this difference was not observed when rays from the sun were substituted for those from the hot body. When a cold instead of a hot body was used the thermometer showed a fall of temperature; hence apparently " cold " could be reflected also. Speculations concerning the nature of this " radiant heat," " liberated fire," " obscure heat " as it was variously termed, were divided between those which represented it as some modified form of light and those which gave it a separate identity; here most theorists suspended judgment. There was, however, a balance of opinion in favour of the view that it was some material emanation rather than vibrations in a " heat fluid " for which some arguments were advanced. That the apparent " reflection of cold " did not revive the seventeenth century dualism is due mainly to Prévost, who in 1791 put forward his celebrated " theory of exchanges." Following de Luc in regarding heat as a discrete fluid whose particles were in constant motion and which was similar to, though different from, light, he held that all parts of space radiated this " perfectly free fire " to one another in amounts which increased with the temperature.

The material theory of heat was thus held to be adequate to deal with all the phenomena known. That it came to be challenged was due less to the demands of empiricists for a satisfactory guiding conception than to certain apparent inconsistencies of which the absence of any indubitable evidence that the " matter of heat " possessed weight was one of the most intriguing. Numerous experiments on this subject were made, culminating in an elaborate series by Count Rumford, which produced an entirely negative result. Rumford's experiments on the production of heat by boring cannon are well known, and with these, published in 1798, and the " weight of heat " investigations, published in 1799, we come to the date at which the Philosophical Magazine took up the task of assisting the prosecution of science. It remains to add only that, contrary to what is often stated, neither Rumford's experiments nor the later hypothetical ones of Davy destroyed the material theory of heat, which persisted far into the nineteenth century until the doctrine of the conservation of energy was able to take over its useful elements and reject the materialistic embroidery.

Light.

In light, the eighteenth century achieved far less than in heat. This was due very largely to the fact that it started with a much greater stock of knowledge, and further advance demanded a correspondingly greater degree of experimental skill. The laws of reflection and refraction, the facts of diffraction and dispersion, the optical characteristics of thin films, and the finite velocity of light were all known in the seventeenth century. Geometry had been applied as far as it would go, and for the outstanding phenomena the mainly corpuscular theory of Newton and the undulatory theory of Huygens offered alternative and at least partly adequate explanations. There was an obvious need for a definite decision between these hypotheses, but the crucial experiments did not become obvious until a century later. All that the eighteenth century could do was to compare the requirements of the hypotheses with facts already known and seek to discard one or the other accordingly. On the purely experimental side the greatest deficiency was in the field of photometry, and some of the most important work of the period lay in the establishment of the foundations of this science.

A discovery of major importance— which, however, was so fundamental that it had no bearing on the purely physical problems of the time—was that of the aberration of light by Bradley in 1725. That this was an optical discovery is, in a sense, an accident. Its significance lies in its relation to motion in general, and that the moving thing which revealed it was light is a consequence of practical rather than theoretical necessities. Clearly, it offered no means of deciding between waves and corpuscles. For that the criterion seemed to be of the same nature as that applicable to the " substance " and " motion " theories of heat. If light were corpuscular its production should result in a gradual wearing away of the luminous body ; whereas there seemed no reason why waves should not go on being emitted for ever. This consideration led the celebrated Euler, for example, to favour the wave-theory since there was no evidence that the sun was being consumed in spite of its exceeding brilliance. True, luminous terrestrial bodies are burnt up, but that Euler attributed to the emission of smoke, not light. Again, if all the stars, as well as the sun, sent out particles in all directions, these particles would interfere with one another and bodies would have hazy outlines. Euler's theory, of course, required an all-pervading ether, but he used this to assist his explanations of magnetism and electricity and gravity also. He attributed differences of colour to differences of frequency of vibration, and amplified his theory in other ways, some of which brought it close to the nineteenth century wave-theory.

The predominant view, however, was the corpuscular one, which could claim the authority of Newton as well as more substantial support. The answer to Euler was given by Priestley. He attempted to estimate the amount of substance lost in the emission of light by measuring the pressure which light exerted on the surface on which it fell. He analysed experiments by Michell in which sunlight was reflected from a two-foot concave mirror on to a copper plate fixed to the end of a horizontal wire which was delicately balanced so as to rotate about a pivot if pressure was applied to the plate. The wire did indeed rotate, and, supposing this to be a genuine effect of light pressure, Priestley was able to calculate from its magnitude that the sun must be losing just over two grains a day, from which, if the density of the sun were equal to that of water, it would follow that its radius must have decreased by about 10 feet since the Creation. This was, of course, an effective answer to one argument of Euler's, though we know now that Priestley had not detected the pressure of light and if he had done so he would not have invalidated the wave-theory. It was one form of that theory, in fact, that made it possible later to estimate how great the pressure of light should be. To meet Euler's "interference" difficulty it was necessary to suppose that the particles of light were so small that their dimensions were negligible in comparison with the distances between them—a hypothesis that, if one adopted the discontinuous view of matter, was consistent with the fact that light could pass through transparent bodies.

Clearly hypothesis was running ahead of experiment in this matter of the nature of light, and more lasting work was done by Bouguer and Lambert in placing the measurement of the intensity of light on a solid basis. To Bouguer we owe the first trustworthy photometers, and to Lambert the foundation of the theory of photometry. The inverse square law was already common knowledge, but Lambert added other principles such as that the eye could judge if two sources of light were equal in brightness or not, but not by how much the brightness of one exceeded that of the other; that "if the same surface is illuminated at one time by m and at another time by n sources of light, each of which has the same intensity and sends its light to that surface in exactly similar circumstances, then the respective degrees of brightness are to each other as m to n"; and that "the brightness of the illumination of a surface falls off in the same proportion as the sine of the angle of inclination of the incident beam to the surface." (The quotations are from Wolf's 'History of Science, Technology and Philosophy in the Eighteenth Century.') Lambert's most thorough treatment of the subject included a discussion also of atmospheric scattering, the intensities of images and other aspects of photometry. He is deservedly commemorated in the name of the unit of brightness.

Although the foundations of the theory of colour had been well laid by Newton, little advance was made in spectroscopy until the nineteenth century. A misconception of Newton's that deviation without dispersion of light was impossible, was corrected independently by Chester Moor Hall in 1733, Euler in 1747 and Dollond in 1758, the first named arguing simply enough that since the lens of the eye produced images without colour, presumably through the combination of the effects of the lens and the surrounding humours, artificial achromatic combinations should be possible. He followed up the argument by making one, but did not publish his discovery. Good refracting telescopes thus again became a possibility, though they were not made on a large scale until much later owing to the difficulty of manufacture of satisfactory flint glass. Phosphorescence created much interest but little knowledge, and one discovery of outstanding but unrealized importance was made by Melvill who, on examining the light emitted from a spirit flame in which various salts were put, noticed that different colours predominated in the different experiments and, in particular, that the bright yellow colour which stood out when sea-salt was used was so sharply distinguished from the surrounding colours that it must be "of one determined degree of refrangibility." Here was the germ of spectrum analysis, but it was not to develop in our period.

On the whole, however, the most significant contribution of the eighteenth century to the science of optics was not in discoveries but in the generation of men from whom discoveries and ideas were later to come. In the year in which the first number of the Philosophical Magazine appeared, Fraunhofer was a boy of 11, Young was 25 but almost unknown, and Fresnel 10. The age which produced these pioneers is not fairly judged by its matured fruits alone.

Magnetism.

Of the various physical phenomena, none bore a closer superficial resemblance to gravitation than magnetism. Both were manifested by their effects on the movements of bodies, both appeared on the small scale and on the large, for Gilbert's hypothesis that the earth was a magnet was universally accepted in the eighteenth century, and both clearly were associated with effects that decreased with increase of distance between the bodies concerned. In spite, therefore, of obvious differences, it was impossible not to suppose that magnetic phenomena were manifestations of forces of the Newtonian type. On the other hand, whereas Newton had specifically limited his endeavours to the formulation of the law of gravitational attraction and had disclaimed all intention of speculating on the *cause* of gravitation, the investigators of magnetism combined research into the law of magnetic force with theories of the nature of magnetism itself. As Newton would probably have told them, the former labours yielded the more valuable dividends.

The original form of the magnet was, of course, the naturally occurring lodestone, but the eighteenth century made considerable progress in the manufacture of artificial magnets, using purely magnetic methods since the connection of magnetism with electricity was unknown. Several analogies between the phenomena indeed were noted, and much speculation was aroused by the curious fact that iron could be magnetized by lightning which, by the middle of the century, Franklin had shown to be an electrical discharge ; but the nature of the relation, and the circumstances necessary for its utilization, were not even suspected. Any method, however, of producing artificial magnets greatly facilitated the investigation of the subject, and although the strongest fields producible were weak when judged by modern standards, they had the outstanding merit of being controllable so that precise measurements could be made.

The idea that the magnetic force varied as the inverse square of the distance from the pole of a magnet must have occurred very early, but its experimental establishment was not easily achieved. Among others, Hauksbee and Musschenbroek sought to derive a general law, but without success. Michell, however, in 1750 was able to show that the most probable explanation of the work of these experimenters and of his own was that the inverse square law was the law of magnetic force. He found that, at the same distance and in the same direction, the two poles of a magnet exerted exactly the same force but in opposite senses, and that this force decreased proportionately to the square of the distance from a pole. With true scientific caution, however, he added: "I do not pretend to lay it down as certain, not having made experiments enough yet, to determine it with sufficient exactness." Others obtained the same result, notably Lambert, who compared the field of a magnet, as plotted with a small compass needle, with the field calculated on the assumption of an inverse square law.

It was Coulomb, however, who made the final indubitable experiments. These were of two distinct kinds, roughly corresponding to the two commonest methods used nowadays in elementary laboratories for determining the moment of a magnet or the intensity of a magnetic field. First, he measured the period of oscillation of a small compass needle under the influence of the earth's field alone and when modified by the presence of a pole of a long thin magnet placed at various distances from it. The possible disturbing factors were taken into account and a definite result arrived at. In the second series of experiments he used a torsion balance of the kind which he had already employed to investigate the law of electrical repulsion. This had in all essentials the same form as the modern instrument, and is too well known to need description. Coulomb's results were published in 1785, and although Gauss was to confirm them later by a still more precise method, the law of magnetic force may be said to have been known from that time onwards.

For a considerable part of the century the origin of magnetism was, strangely enough, generally attributed to Cartesian vortices which, so far as gravitation was concerned, had been thoroughly discredited by Newton's achievements. Euler pictured the vortices as passing through pores in the magnetic body, but always in one direction only, thus producing opposite polarities at entrance and exit. Later, there came "one-fluid" and "two-fluid" theories, the former having the advantage of simplicity in the explanation of either attraction or repulsion, but becoming decidedly embarrassed when both had to be accounted for. Even a two-fluid theory did not readily explain the *inseparability* of the magnetic poles, and it was probably this awkward fact that demanded recourse to the discontinuous theory of matter—a hypothesis which seems in the eighteenth century to have been called upon when needed and at other times ignored. The ultimate particles of a potential magnet were conceived as being themselves magnets, and the process of magnetization consisted simply in their alignment. This view, which has a quite modern appearance, has in fact a very respectable antiquity. It has been suggested that Gilbert may have held it, though the evidence is not convincing, but it was certainly no stranger to eighteenth century thought. Among others, Swedenborg, applying his own theoretical ideas to the experiments of Musschenbroek, gave an account of molecular magnetism with diagrams that might find a place in modern text-books. The remarkable work of this philosopher, in this as in other fields, had little influence on the general progress of science, but the molecular theory of magnetism found independent expositors who could claim a considerable following.

The phenomenon of magnetism is often classified today under three headings—ferro-, para- and dia-magnetism. An example of the first of these was known to the early Greeks ; the others first came to light in the eighteenth century. Cobalt and nickel were discovered during this period and shown by their discoverers to possess relatively feeble magnetic properties. Coulomb made an extended investigation of the possibility of magnetization, and added other substances to the list. The more startling fact of diamagnetism was discovered by S. Brugman in 1778 : he found that both bismuth and antimony repelled the poles of a magnetic needle. The discovery seems to have been without influence on theories of magnetism at this time.

The chief contributions to the subject of terrestrial magnetism lay in the accumulation of data concerning the variation (as declination was then called) and dip in different parts of the earth, and the diurnal and annual changes in the variation at one place. Wilcke, in 1768, published the earliest chart of a considerable region of the earth's surface (including most of the earth except Asia and North America) which showed isoclinic lines. This was a considerable achievement, though the accuracy was admittedly not of a high order. A variation chart had been published by Halley at the very beginning of the century and was frequently revised thereafter. Towards the end of the century, the studies of distribution of the magnetic elements, which had previously been mostly confined to variation and dip, were extended to the intensity of the earth's field. The measures were relative, the first absolute determination of magnetic intensity being reserved for Gauss in the next century. Speculation on the origin of terrestrial magnetism, though not entirely non-existent, was fortunately restrained in view of the scantiness of the data.

Electricity.

Of all the phenomena with which physics is concerned, there is none to which the contribution of the eighteenth century is so fundamental as it is to electricity. Heat, light and sound, and even magnetism, were familiar enough before ; one could hardly escape them in ordinary, everyday experience. But electricity was not normally encountered. To observe it special things had to be done, and the results attending these were so slight as to escape the notice of all but the most curious. The subject was not completely unknown ; the ancient Greeks, of course, had observed the peculiar property of rubbed amber, and Gilbert and von Guericke during the previous century had given some attention to it, but beyond the construction of a simple frictional machine and the recognition that there was a distinction between what we now call conductors and non-conductors, we may say that in the year 1700, complete ignorance existed of what was later to become the most comprehensive department of physical knowledge.

Throughout the century, the basic phenomena progressively came to light—the basic phenomena, that is, of what we now call electrostatics. A current of electricity was unknown. The eighteenth century had to discover the various manifestations of electricity, to prove that they were manifestations of the same thing, to devise instruments for their measurement, and to form concepts for their understanding. Its success was very great and very hard won. A striking feature of its attainment was the fearlessness with which the investigators used themselves as instruments of detection, despite the unknown dangers lurking in this strange effluvium. The experience of a shock seemed to be the standard evidence for the presence of electricity, and even when sufficient knowledge had been obtained to adopt safer methods, it continued to be used. Musschenbroek, reporting to Réaumur his discovery of the Leyden jar in 1746, wrote : " I wish to report to you a new but terrible experiment, which I advise you on no account to attempt yourself." He had, however, evidently repeated it a number of times, for he described its dependence or otherwise on the shape and material of the vessel used. Musschenbroek's letter was read to the French Academy, whereupon Nollet and Le Monnier immediately repeated the experiment. Richmann, whose views on heat we have already considered, was killed in 1753 while experimenting on the electrical nature of lightning. Nevertheless, the electric shock continued to be regarded as almost an essential part of an electrical experiment.

Early electrical machines, of which Hauksbee's in the first decade of the century is an elementary type, consisted of a rotating body, usually glass, rubbing against the human hand or some other material. The charge was collected by a " prime conductor " and transferred where needed. Such instruments, however, did not become generally used until halfway through the century, and in Gray's pioneer experiments on electrical conduction, which were made about 1730, only the simple manual operation of rubbing was used. Gray discovered that when one part of a system was rubbed, not only that part but a distant unrubbed part could also attract light bodies. This suggested that the electrical influence was being conducted from one place to another, and by an extension of his experiments, Gray was able to conduct it over a distance of some 800 feet. Difficulties with the supporting materials for such long conductors revealed the existence of two classes of bodies, which Desaguliers later called *conductors* and *electrics per se* and which evidently corresponded to what Gilbert had called non-electrics and electrics.

The discovery that there were two kinds of electricity as well as two kinds of electrifiable bodies was made by du Fay shortly afterwards. He found first that an electrified body attracts non-electrified bodies, and after communicating electricity to them, repels them ; and secondly, that there are two opposite kinds of electricity which he called *vitreous* and *resinous* (the names *positive* and *negative* were given later by Franklin, and indicate the transition from a purely descriptive to a theoretical account of the phenomena). Neither terminology is satisfactory, as Wilcke and Canton independently found about 1754. The same body, whether " vitreous " or " resinous " in character, may develop either kind of electricity according to the material with which it is rubbed ; and the difference between the two kinds is not that between excess and defect as Franklin supposed).

The introduction into electrical experiments of the Leyden jar, which was soon, in battery form, to become the chief instrument of research with what were regarded as large charges of electricity, gave a great impetus to the study of the science. This was due independently to von Kleist of Pomerania and Musschenbroek of Leyden, who almost simultaneously made the discovery accidentally. It led, among other things, to attempts to discover the speed of propagation of the electric fluid. Several observers discharged Leyden jars over great distances, but no appreciable interval was noticed between the shocks or sparks observed at widely separated points. Franklin was openly sceptical of the value of such experiments. Thinking of a discharge as a movement of a fluid which filled the whole circuit, he held in effect that the interval between the passage of electricity across two separate points of the circuit no more indicated the velocity of electricity than the interval between pushing one end of a rod and the movement of the other end would indicate the speed of movement of the rod.

Franklin's best known contribution to electricity is his famous demonstration of the electrical nature of lightning by means of his kite. This experiment was performed in 1752, but he had already, in 1749, set out the evidence, in the form of a number of analogies, for his belief in this identification, and the kite experiment was rather a confirmation of a belief already held than an original source of conviction. The kite, made of silk on a cedar-wood frame, was flown when a thunderstorm seemed to be approaching, and to the end of the cord near the hand was attached a piece of silk which was to be kept dry and by which the apparatus was held. A key was placed at the junction of the cord and silk, and when the kite and cord were electrified by an approaching thundercloud, electricity could be drawn off from the key and shown to possess all the properties of laboratory electricity. By repeated experiments, Franklin was able to show that thunderclouds were sometimes positively and sometimes negatively charged.

Franklin's experiment was in essence performed earlier under the supervision of Dalibard at Marly, near Paris. Dalibard had been impressed by Franklin's arguments, and sought a direct proof, using a pointed iron rod about 40 feet high instead of a kite. Having to be absent during the thunderstorm, he left the observations to Coiffier, an ex-dragoon, and Raulet, the Curate of Marly. Sparks about 1½ in. long were obtained at intervals which Raulet estimated at between a Pater and an Ave (freedom of choice of time scales is not a new idea). Raulet received the spark in person, sustaining a wounded arm and a charge of having acquired a smell of sulphur, but the theological implications of the accusation were not established.

The practical application of his discovery for the protection of buildings and ships soon occurred to Franklin, and as early as 1750 he described the principle of the lightning conductor. It was immediately taken up and some thirty years later became the occasion of a controversy which, in its confusion of scientific facts with political doctrines, seems more characteristic of twentieth century Europe than eighteenth century England and America. Franklin had prescribed pointed upper ends for lightning conductors, but Wilson and others thought that these would attract lightning that might otherwise take another course and that the ends should therefore be blunt. The ensuing controversy dominated the contributions on electricity to the 'Philosophical Transactions' for a considerable period, and was said to have led to the resignation of Sir John Pringle from the Presidency of the Royal Society. The following passage appears in the biography of that gentleman in the 'English Cyclopaedia':—

"About the year 1778 a dispute arose among the members of the Royal Society relative to the form which should be given to electrical conductors so as to render them most efficacious in protecting buildings from the destructive effects of lightning. Franklin had previously recommended the use of points, and the propriety of this recommendation had been acknowledged and sanctioned by the Society at large. But, after the breaking out of the American Revolution, Franklin was no longer regarded by many of the members in any other light than an enemy of England, and, as such, it appears to have been repugnant to their feelings to act otherwise than in disparagement of his scientific discoveries. Among this number was their patron George III, who, according to a story current at the time, and of the substantial truth of which there is no doubt, on its being proposed to substitute knobs instead of points, requested that Si John Pringle would likewise advocate their introduction. The latter hinted that the laws of nature were unalterable at royal pleasure; whereupon it was intimated to him that a President of the Royal Society entertaining such an opinion ought to resign, and he resigned accordingly." 'The Dictionary of National Biography,' however, makes no reference to the controversy and merely says that Pringle felt compelled to resign after his health began to fail.

All the work so far described was purely qualitative—an indication of the backwardness of the subject of electricity compared with heat, say, where we were able to start with the thermometer already in existence and the question of giving a meaning to its readings ripe for consideration. Even the first primitive electroscopes did not come into use until towards the middle of the century. They usually took the form of suspended threads, with or without pith balls at the ends. The next stage in precision was the control of the discharge so that it did not exceed the danger limit—a precaution especially necessary after the discovery of the Leyden jar and the use of electrical treatment in medicine. This was achieved by including a gap in the circuit, the length of which determined how strongly the jar could be charged before discharging through the patient. The earliest actual measuring instruments were simply electroscopes with a scale of degrees attached to indicate the angle of separation of the threads. Henly's "quadrant electrometer" (not to be confused with the modern instrument of that name), produced in 1770, was of this type. Volta, who had created a valuable instrument in the *electrophorus*, by which successive, presumably equal, quantities of electricity could be transferred from a cake of resin to a conductor, wished to establish an absolute scale of electrical measurements, and it had already been suggested that this could be achieved by balancing the attraction of a charged balance pan by weights placed in the other pan. Instruments of this sort were, in fact, made, but the unit of electric charge had to be defined before the desired result could be achieved, and the existing ideas of electricity were too undeveloped for this problem to be satisfactorily solved.

The variation of electrical attraction or repulsion with distance, however, was a soluble problem, irrespective of the measurement of electricity itself. The inverse square law was, of course, suspected long before it was established, and experiments indicating that it was the actual law of electric force were not wanting before Coulomb's definitive demonstration. Thus Priestley, in 1766, showed the absence of electric force from the inside of a charged vessel, and suggested the inference that the law of attraction must be the same as that of gravitation; since it was known that if the Earth were a shell, a body inside it would not be attracted to one side or another. Cavendish followed up this work with much more precise experiments of the same kind, in which he proved that the charge on a conductor lay entirely on the surface. From the delicacy of his experiments he concluded that the electric force varied inversely as some power of the distance lying between $2+\frac{1}{50}$ and $2-\frac{1}{50}$.

Coulomb's investigations with the torsion balance were carried out from 1784 onwards. As with magnetism, two types of experiment were performed. For electrical repulsion the force was measured by the angle through which the torsion head had to be turned to maintain a chosen distance between the electrified balls. For attraction a small charged body was oscillated in the neighbourhood of an oppositely charged sphere, and its frequency was noted. In these experiments, of course, the (then immeasurable) magnitude of the charge did not matter so long as it remained constant. Coulomb investigated the leakage of charge and made corrections for it. He also obtained Cavendish's result concerning the distribution of charge over the surface of conductors, and realized that this followed from the inverse square law.

The introduction of measurement into electrostatics inevitably raised questions about the nature of the thing measured. All theories supposed electricity to be some sort of substance, so that the idea of quantity of electricity as the object of measurement was a natural one. It was not, however, adopted automatically. Thus, Priestley, in his 'History of Electricity,' first published in 1767, wrote: "By *electricity* I would be understood to mean, only those *effects* which will be called electrical; or else the *unknown cause* of those effects"—indicating that he was prepared to admit at least the possibility that electricity was an intensive, rather than an extensive, quantity. Cavendish went further and likened the "degree of electrification" to the pressure in a fluid. This "degree of electrification" determined in which direction electricity would flow when two charged conductors were connected with one another. He was evidently approaching the idea which later came to be known as " potential " and has played so important a part in electrical theory. He also conceived the modern notion of "capacity," which he measured in " inches of electricity," and to some extent anticipated Faraday's work on specific inductive capacity. All this is very remarkable at so early a stage, when both conceptions of electricity and measuring instruments were of the crudest. True to his character, he kept most of his results to himself, but it may be doubted whether ideas so far in advance of their time would have had much general effect had they been published.

Another modern conception which was foreshadowed in the eighteenth century was that of electrical resistance. Aepinus, about 1759, conceived that the difference between conductors and non-conductors was simply a matter of degree, the former offering less opposition to the passage of electricity than the latter. Priestley some ten years later showed that the discharge of a battery of jars was independent of the shape of the external circuit, but did depend on its length and cross-section. His criterion for the intensity of the discharge was the length of spark gap that would be bridged in preference to a fixed alternative circuit. Cavendish extended the experiment to solutions of various strengths and studied the effect of temperature on resistance. He showed, further, that the resistance of a conductor did not depend on the strength of the discharge, and described how the discharge was distributed among several conductors in parallel. We are so accustomed to associating these ideas with the phenomena of steady electric *currents* that we are apt to overlook how much it is possible to learn from discharges of static electricity. Cavendish, however, was a genius of the highest order, who, but for his extreme secretiveness, would figure much more prominently in the history of science than he does.

All this work was of the nature of prophecy rather than of principles for immediate guidance ; it awaited its Faraday for its proper understanding. So far as working hypotheses were concerned, the eighteenth century, in electricity as in everything else, looked to effluvia, or fluids, or ethers for the explanation of the curious things it was bringing to light. Since the phenomena of magnetism and electricity were so similar it was natural that explanations of the same type should be called forth. There were supporters of both one-fluid and two-fluid theories, the general arguments for their reality and descriptions of their modes of application being similar to, though by no means entirely identical with, the corresponding accounts in magnetism. There was, it is true, the simplifying feature that the two kinds of electrification, unlike the two kinds of magnetism, were not inseparable, though Wilcke showed in 1757 that in electrification by friction both were invariably produced. There was also a general idea that a charged body was surrounded by an " electrical atmosphere "— a notion which offered a ready, if only partial, explanation of electrification by induction which was discovered by Canton and studied more completely by Wilcke and Aepinus about the middle of the century. But something more definite was required by the intellectually curious, and it was inevitable that a variety of views should be held, according to the particular phenomena among the growing number that were considered to be most evidential.

Euler, whose luminiferous ether has already been mentioned, gave what is probably the first electromagnetic theory of light by making his one all-pervading ether serve for all physical phenomena that seemed to require such a medium. If a body had ether forced into it, it acquired one kind of electrification ; if it suffered a loss from its normal quantity it acquired the other kind. He evidently wished to avoid the multiplicity of ethers which threatened to drag science back to the "Aristotelian" philosophy of "occult qualities" against which Galileo and Newton had contended. Franklin, confining his attention to electricity, also postulated an all-pervading ether, an excess of which gave a body a positive charge while a defect gave it a negative charge. This ether was a sort of matter, but its particles were extremely fine and did not attract but repelled one another. Ordinary matter and electrical matter were mutually attractive, and a material body normally gathered within itself as much electrical matter as it could accommodate, so that when more was added it lay on the surface and formed an atmosphere around the body. This theory was regarded favourably, though without enthusiasm, by Priestley. A somewhat similar theory was held by Aepinus and Cavendish, with, however, one important difference ; namely, that the particles of ordinary matter, as well as those of electricity, repelled one another, the law of repulsion being the same for both and involving a force varying inversely as some power of the distance less than the cube. The body was thus held together by the attraction between its material and its electrical constituents. Substitute for matter, atomic nuclei, and for electricity, electrons, and this theory becomes identical with that of the present day. This was a remarkably bold hypothesis to hold in view of the known gravitational attraction of ordinary uncharged matter.

The weight of authoritative opinion thus seemed to favour a one-fluid theory, but the other view was very much alive also. It had the advantage of the one-fluid theory in the explanation of electrical repulsion, and it was this phenomenon that led du Fay to introduce it on his discovery of the difference between "vitreous" and "resinous" electricity. Nollet, who in his early days collaborated with du Fay, seems to have held to a single fluid, but provided bodies with two sets of pores through one of which electricity could enter bodies while through the other it could leave them. As with heat, much labour was spent in an attempt to discover the weight of electricity, and since no success was obtained, theorists were at liberty to suppose the weight positive or negative—a great advantage in dealing with awkward phenomena. On the whole it might be said that neither theory led to any outstanding discovery which could not have been made by means of the other, and the various views, though doubtless serving to coordinate the facts in the minds of the experimenters, could claim little other intrinsic merit.

It was reserved for the last years of the century to reveal the facts which were to give an entirely new turn to the subject. In 1780 Galvani noticed that a frog's leg became agitated in the neighbourhood of an electrical discharge. This small beginning led him to further experiments culminating in the observation that similar convulsions could be produced in frogs attached to a brass hook and an iron trellis when the metals came into contact. The agitation seemed to require no special external conditions for its occurrence, and this made it clear that here was no manifestation of atmospheric electricity or any other known electrical action, but something hitherto unknown. Galvani ascribed it to " animal electricity," a property peculiar to organic bodies, but Volta—with, as it turned out, truer insight—thought the contact of the metals was the source of the electricity, the frog serving merely to show it. Acting on this idea—to which he was converted from an original agreement with Galvani's hypothesis—he proceeded through a series of experiments to his fundamental discovery of electrification of two different metals by contact, without the presence of any third substance. This he described in 1797, and in such an atmosphere of promise the ' Philosophical Magazine ' in the following year drew its first breath.

Two things strike us in comparing the scientific work of this century with that of our own. In the first place the sciences were completely isolated from one another. Each was still very close to its basic phenomena and had not advanced far enough to encounter the others. Hence the tendency to multiply fluids, ethers, forces and what not. This was not a fault but a sign of immaturity. The facts had not been discovered which would have justified the use of the same concept for a variety of purposes, and such a unifying conception as, for instance, the ether of Euler, was not of any real value. Much more efficacious for further advances were the separate hypotheses which dealt more completely each with its own branch of study. The day had not arrived when such sciences as electro-magnetism, thermo-dynamics, photo-chemistry and so on could profitably be conceived.

Secondly, despite the separation of the sciences, the same men studied them all. The gradual unification of physics has paradoxically been attended not by a diminution but by an increase of specialization. In reviewing the various branches of physics, chemistry, astronomy and mathematics in the eighteenth century we find the same names occurring everywhere—Priestley, Cavendish, Wilcke, Michell, Lambert, Euler, to mention only a few. It was the spring-time of modern physics, when something new appeared every day and the man who had a mind for such things could take his pleasure where he would. That this is impossible today is doubtless a necessary result of progress, to be welcomed rather than deplored. Yet, looking back on an age now gone for ever, one might ask pardon for a faint feeling of regret.

CHEMISTRY THROUGH THE EIGHTEENTH CENTURY

By Prof. J. R. PARTINGTON, D.Sc.

In the year 1700 Chemistry had gone a long way towards becoming a science. It had become enriched by the acquisition of a large body of experimental knowledge, and serious attempts had been made to reduce this to some kind of system. The principles underlying this systematization were, it is true, often incorrect, but they were conceived in terms of contemporary scientific theory and were very different from the views of the alchemical school a century before. The rationalization of outlook owed much to the active progress in mathematical astronomy and physics which followed the work of Newton, and his influence may be traced in some parts of chemistry during the whole of the eighteenth century. Interesting views were put forward in Newton's letter to Boyle [1] in 1679, and more particularly in the "Queries" in the later editions of the 'Opticks.' The latter, although containing some important material previously published by Mayow and by Stahl, introduced the idea that chemical changes might be explained in terms of atoms with short-range attractive forces acting between them, and suggested that chemistry might turn out to be a branch of mechanical philosophy. This view was the foundation of a theory proposed by John Keill [2], which assumed a power in matter superadded to gravity, but diminishing in a greater ratio than the inverse square of the distance, and an attempt was made to explain the phenomena of cohesion, elasticity, crystallization, solution, fermentation, effervescence, etc. The Newtonian theory was also used by John Freind in his Oxford lectures [3] to explain the solution of metals in acids, algebraical relations being given. The construction of tables of affinity was attempted by E. F. Geoffroy [4], but the antagonism of the French Academy to Newton's ideas caused him to replace the word "attraction" by the neutral term "rapport," which, in fact, expressed the old view that affinity acts most strongly between substances having the greatest relation to one another. These affinity tables persisted through the century, finding their most elaborate form in Bergman's tables (1775) [5]. They assumed that affinities are absolute forces; if the attraction of a substance A for another B is greater than the attraction of A to a third substance C, then the substance B will act upon a compound AC so that the change $B+AC=AB+C$ takes place completely. The tables are the result of chemical experiments, and no use was made of any theory of attractive forces. That the results could be different in solutions and in dry reactions, and that a large excess of a substance is sometimes necessary to cause a reaction to go to completion, did not escape Bergman's notice, but the chemical theory of affinity had to await the acute criticisms and ingenious experiments of Berthollet, about 1800, before the idea of the action of mass was introduced. Berthollet's views were also founded on Newton's, but again the actual results were based on purely chemical experiments.

Boyle and Newton made much use of the atomic theory, which seems to have come to them from the old Greek source of Epicurus by way of Gassend [6] In the form in which Newton had left it, the atomic theory was unsuitable for later chemical applications, since air in Newton's time was still an element, and he could not, therefore, discuss the problem of the structure of mixed gases, the intensive study of which led Dalton to his chemical atomic theory [7].

Of great influence in the early eighteenth century was the famous text-book 'Elementa Chemiae' (1732) of Hermann Boerhaave, professor of chemistry at Leyden from 1718 to 1729. Boerhaave's definition of chemistry was given (long after it had become antiquated) in the 'Dictionary' of Samuel Johnson (who greatly admired Boerhaave, and wrote a biography of him): it would apply equally to cookery :—

> 'Chemistry is an art which teaches the manner of performing certain physical operations, whereby bodies cognizable to the senses, or capable of being rendered cognizable, and of being contained in vessels, are so changed by means of proper instruments, as to produce certain determinate effects; and at the same time discover the causes thereof, for the service of various arts'.

The first volume of the book, dealing with theory, neglects Stahl's phlogiston theory (see below), although Boerhaave's own theory, that combustibles contain a "food of fire" (*pabulum ignis*), is very like Stahl's. He describes experiments of Boyle, Du Clos, and Homberg on the increase in weight of metals on calcination, and he proved that the matter of fire has no weight by showing that a mass of 8 lb. of iron has the same weight red-hot and cold.

Parts of the chapter on air are in contradiction, some having been written before the appearance of Hales's 'Vegetable Staticks' in 1727; Boerhaave adopted to some extent the doctrine of the fixation of air in bodies, but (perhaps because he did not think this sufficiently proved) he did not retract what he had previously said to the contrary.

In a section on menstruums, he says the solvent action depends on motion, the small particles of the menstruum strongly cohering with the particles of the solute (*arctissime adhaerentium ad corpusculi soluti per Menstruum corporis*); it is not easy to understand the physical manner by which this motion is excited, but it seems that it is not due to impulse, gravity, elasticity, magnetism, etc., but to a particular cause, not common to all bodies, exerted between the solvent and solute (*singularis enim est inter solvens et solvendum, non communis omnibus corporibus*). Boerhaave, although using mechanical analogies, and quoting Newton, did not think chemical changes could be explained mechanically, but thought they depended on some other cause.

The name gas, and the recognition that there are several gases different in properties from atmospheric air, appear in the writings of Van Helmont

(1648) in the preceding century [8]. Van Helmont, although sharing the credulity of his time and under the influence of alchemy, marks a transition to Boyle's outlook, because he sought rational explanations, carried out quantitative experiments, and taught that substances preserve their individuality in chemical changes ; a metal dissolved in an acid, for example, is recoverable from the solution. Van Helmont's rationalism, although not so complete as they would have wished, appealed to Boyle and Newton, the former frequently referring to him with approbation. Alchemy had had an independent philosophy, which was used by Francis Bacon, and in its rationalized form with marked success during the seventeenth century by Boyle, Tachenius, Glauber, and Kunckel. This method, which will be called chemical to distinguish it from the mathematical or quantitative method of astronomy and physics, was to be used with brilliant results by Scheele and Priestley. Whatever views Newton may have held in private on alchemy, they do not intrude in his published works, on which alone his attitude to chemistry should be judged [9]. He had a good knowledge of the new method in chemistry, and did not, as some of his followers did, believe that chemistry could make progress by the use of the mathematical method alone.

Boyle, and especially Mayow (1674), had collected some gases and the latter had shown that they had the same compressibility as common air, but Boyle called them "factitious airs," or "air generated *de novo*," and the existence of several gases with chemical properties different from those of common air was first clearly recognized in the second half of the eighteenth century. Stephen Hales, in his 'Vegetable Staticks' (1727), had collected several gases but although he paid great attention to the quantities of these "airs" obtained from given weights of different materials, following the quantitative method, he completely neglected their chemical properties and reached very faulty conclusions. He says little more than that "air abounds in animal, vegetable and mineral substances," and that it is :—

> 'very instrumental in the production of the growth of animals and vegetables, both by invigorating their several juices, while in an elastick active state, and also by greatly contributing in a fix'd state to the union and firm connection of the several constituent parts of bodies, viz. their water, salt, sulphur and earth.'

Hales's apparatus was clumsy, but his work is important in having attracted Cavendish and Priestley to the study of gases.

The earlier issues of the Philosophical Transactions contain several papers on "airs" and "damps," mostly unimportant. A notable paper is that of William Brownrigg (1765) on "An Experimental Enquiry into the Mineral Elastick Spirit, or Air, contained in Spa Water." This had been read twenty-four years before it was published, and was known to Hales. It deals with carbon dioxide, but the chemical properties are not adequately described. A primitive form of pneumatic trough, with a shelf above the water in which bottles were secured by wedges, was used.

Publication on gases really began with Cavendish's paper [10] in 1766. It describes the properties of two known gases, fixed air (carbon dioxide) and inflammable air (hydrogen), the determinations of their densities, and the solubility of fixed air in water and spirit of wine. The properties of hydrogen were first clearly described by Cavendish, who is sometimes regarded as its discoverer. Cavendish collected and stored fixed air over mercury, a method later much used by Priestley. He also recognized (1766, 1788) the existence of equivalent weights of reacting substances and used the name equivalent. Some important purely chemical work he did not publish. He did not discover any gas not previously known. Inflammable air had been collected and examined by Boyle, and fixed air (already recognized by Van Helmont) had been shown to play a part in chemical changes by Black.

Joseph Black was born on 16th April, 1728, at Bordeaux, where his father, of Scots descent, but born in Belfast, had a wine business. Joseph was educated in medicine at Glasgow University (1746), where he came under the influence of Dr. William Cullen (1710–1790), then lecturer in Chemistry, and worked in Cullen's laboratory. He succeeded Cullen in 1756 as lecturer in chemistry at Glasgow, and in 1766 as professor of chemistry in Edinburgh, occupying the last position until his peaceful death, seated in his chair, on 6th December, 1799. Black was one of the most notable teachers of chemistry in the century ; ill-health prevented him from prosecuting much original research, and apart from the work to be mentioned he published practically nothing. His lectures were taken down by students, and manuscript copies are still available ; they were published by his pupil Robison in 1803, and German translations appeared in 1804–5 and 1818.

Black's systematic and critical mind, and his wide knowledge of chemistry, enabled him to form correct judgments of new discoveries, and he was one of the first anywhere to accept Lavoisier's new theory. His influence in Edinburgh was very great, and led to a keen interest in Lavoisier's work, the 'Traité Élémentaire de Chimie' (1789) of the latter being translated and published in Edinburgh in 1790. Black had accepted the new theory at least as early as 1784.

Black's inaugural dissertation for the M.D., 'De Humore Acido a Cibis Orto, et Magnesia Alba,' was presented in 1754, and an extended version of the chemical part was published [11] in 1756. In this he showed by careful quantitative experiments that "mild" magnesia, limestone, and the so-called "mild" alkalis, contain what he called *fixed air* (carbon dioxide), which was the same as Van Helmont's *gas sylvestre*. Calcined magnesia, quicklime, and caustic alkalis contain no fixed air, and owe their caustic properties to the deprivation of the mild forms of the fixed air contained in them. The mild substances effervesce with acids because the acid has a greater affinity for the caustic form than the fixed air has, and hence it displaces the fixed air, which escapes in the elastic state.

Black's quantitative work marks a turning point in chemical method. He was not the first to use the balance in chemistry, but he drew far-reaching theoretical conclusions from its use. These conclusions, simple as they now seem, were not obvious ; in 1764 Johann Friedrich Meyer, an apothecary of Osnabrück, published a book in which caustic alkalis and quicklime were alleged to contain an otherwise unknown fiery acid, *acidum pingue* (giving their solutions a soapy feel), and since they were already saturated with this acid, they did not effervesce with other acids. Meyer's theory was taken seriously by Lavoisier in 1772, and there was a prolonged controversy between the supporters of Black and of Meyer.

Meyer's *acidum pingue* was like, if not identical with, another figment of the imagination. Johann Joachim Becher (1635–1682) in his ' Actorum Laboratorii Monacensis, seu Physicae Subterraneae ' (1669) stated that the constituents of bodies are air, water, and three earths, the inflammable (*terra pinguis*), mercurial, and vitreous. The *terra pinguis*, regarded as a constituent of combustible bodies, was re-named *phlogiston* by the famous Georg Ernst Stahl (1660–1734), from 1716 physician to the King of Prussia in Berlin. He gave a sketch of the phlogiston theory in his small ' Zymotechnia Fundamentalis ' (1697), based on his lectures in Jena, and expanded it in a supplement (' Specimen Beccherianum ') to his edition of Becher's book (1703) and in two versions of a text-book of chemistry, ' Fundamenta Chemiae ' (1723 and 1746–7), a somewhat altered English version of the 1723 issue of the latter being published in 1730.

According to Stahl, combustible bodies and metals contain a principle of inflammability, phlogiston, which is identical in them all and can be transferred from one body to another. When bodies burn, or metals are calcined, phlogiston escapes. When the calx of a metal is heated with charcoal or oil, phlogiston is restored and the metal is formed. A calx is an element, a metal is a compound of its particular calx and phlogiston ; sulphur is a compound of sulphuric acid (an element) and phlogiston. The last proposition was proved by showing that when the compound of sulphuric acid and potash is heated with charcoal (rich in phlogiston) the same liver of sulphur (potassium sulphide) is formed as is obtained on fusing potash with sulphur :—

$$\text{(sulphuric acid+potash)}+\phi=\text{liver of sulphur,}$$
$$\text{sulphur+potash}=\text{liver of sulphur,}$$
$$\therefore\quad \text{sulphuric acid}+\phi=\text{sulphur.}$$

Stahl recognized the existence of sulphurous acid intermediate between sulphur and sulphuric acid, and, following Mayow (1674), he gave clear examples of the displacement of one metal from its salt by another metal, so arriving at an affinity series which is reproduced by Newton.

Stahl's definition of Chemistry is much better than Boerhaave's :—

> ' Chemistry, or Alchemy, or the Spagiric Art, is the art of resolving mixed, compound, or aggregated bodies into their principles, and of composing such bodies from those principles.'

After the work of Black and Cavendish, the chemistry of gases was greatly advanced by the researches of Scheele and Priestley, carried out independently about the same time, and both leading to the discovery of oxygen.

The great extension of chemical knowledge represented by the discoveries of gases by Scheele and Priestley was the result of following the chemical experimental method. They made little use of the quantitative procedure which, too rigidly followed, had sometimes proved more of a hindrance than a help. Scheele and Priestley, by qualitative methods, made discoveries from which a new science of chemistry could be created by Lavoisier by the use of the quantitative method, but Lavoisier made no discoveries of new substances and, without the facts made available to him, there would have been no new quantitative science.

Carl Wilhelm Scheele, born in 1742 at Stralsund, then the chief town of Swedish Pomerania, was an apothecary. Throughout his life he was in modest circumstances and his work in Malmö, Stockholm, Uppsala, and Köping, was carried out in very unfavourable conditions which accelerated his early death in 1786. Scheele belongs to the few really great chemists, and as Hoefer said: "avec de petites ressources, il fit de grandes choses." His steady devotion to chemistry, undeflected alike by disappointment or success, was the guiding star in his short life. In a letter to his friend Gahn he says: "How glad is the enquirer when discovery rewards his diligence; then his heart rejoices," and if Scheele's reward in discoveries was rich, it was abundantly deserved. In recognizing the early genius of Scheele, giving him encouragement, and helping to bring his discoveries to the notice of contemporary chemists, great credit is due to Bergman, then the eminent professor of chemistry at Uppsala.

Scheele's researches began before 1770; those leading to the discovery of oxygen seem to have been made about 1772, if not earlier [(12)]. They were announced in a book, 'Chemische Abhandlung von der Luft und dem Feuer,' which was sent to the printer in 1775 but did not appear until 1777, when oxygen had been independently discovered by Priestley, and Lavoisier was just announcing his new theory. Scheele's laboratory journal has been preserved, and his priority is without question. Priority in discovery need not depend on date of publication if written records of unquestionable date are available.

Although Scheele's discovery of oxygen is best known, his other discoveries are no less notable. They include chlorine, manganese and barium (in its compounds) (1774), molybdenum (1778) and tungsten (1781) as oxides, arsenic hydride and arsenic acid (1775), silicon fluoride and hydrofluoric acid (1771 and 1781), copper arsenite (1775–8), and a large number of organic compounds: tartaric and pyrotartaric (1770), oxalic (1784), mucic (1780), lactic (1786), uric (1776), prussic (1782–3), citric (1784), malic (1785), gallic and pyrogallic (1786), acids, glycerol (1783–4), murexide (1776), several esters and aldehyde (1782), and casein (1780)

About 1772, if not earlier, Scheele had concluded that air is a mixture of two gases, fire air which supports combustion and respiration, and foul air which does not. He had extracted fire air, which is oxygen, from nitric acid, nitre, mercuric oxide, manganese dioxide, and silver carbonate, discovered its most important chemical properties, and showed that it is rather heavier than air. This involved some quantitative work. Scheele was guided in these researches by a peculiar form of the phlogiston theory. He assumed that heat is a compound of fire air and phlogiston (ϕ) : heat=fire air+ϕ. If a substance presented to heat has a greater attraction for phlogiston than fire air has, such as red precipitate of mercury or calx of mercury (mercuric oxide), this will rob the heat of phlogiston and set free the fire air :

$$\underbrace{(\text{fire air}+\phi)}_{\text{heat}}+\text{calx of mercury}=\underbrace{(\text{calx of mercury}+\phi)}_{\text{mercury}}+\text{fire air.}$$

Scheele's views on phlogiston (which he called the "inflammable principle ") were not quite consistent ; in some cases he identified it with inflammable air (hydrogen), as had been done by Cavendish and Kirwan, and when he discovered chlorine by heating marine acid (hydrochloric acid) with pyrolusite (manganese dioxide), he regarded the product as (marine acid$-\phi$)=(marine acid$-$hydrogen), which Davy afterwards showed is correct. Lavoisier and the French school, misled by a theory that all acids contain oxygen, regarded chlorine as oxidized marine (or muriatic) acid.

Joseph Priestley [13] was born near Leeds in 1733 and died, a refugee from political persecution, in America in 1804. His main interest was theology, in which he was heterodox, but he became interested in science and, as literary companion to Lord Shelburne and later as a friend of Wedgwood, he was able to carry out his experimental work in easy circumstances. His chemical experiments began about 1767 with fixed air formed in vats in a brewery, and he continued work with gases, becoming acquainted with the earlier work of Hales and Cavendish, which served as a model for his own. His apparatus and technique were much neater than theirs, and his trough with a shelf for collecting gases over water, and his use of mercury for collecting soluble gases (already applied by Cavendish in 1766), enabled him to discover many new gases. Priestley's method was largely but not entirely qualitative, and by pressing on with such work he acquired a technique in the manipulation of gases which made him an acknowledged expert in this difficult field. He was little interested in theory, some rather crude forms of the phlogiston theory, which he changed as he thought fit, sufficed for his needs, and he was, in consequence, quite unable to draw correct conclusions from his results. He can be blamed for his obstinate retention of the phlogiston theory, since his views were taken seriously by other chemists. Much time and effort expended in traversing Priestley's erroneous theories could have been better used in other directions.

In March 1772, Priestley read to the Royal Society a paper " Observations on Different Kinds of Air." This was later published in the Philosophical Transactions, with additions made up to the end of 1772, and describes the discovery of nitrous air (nitric oxide, discovered in June, 1772), acid air (hydrogen chloride), and (less definitely) nitrogen, and carbon monoxide. The confusion of the last gas with hydrogen was the source of most of his mistakes. His later work is largely contained in his book, ' Experiments and Observations on Different Kinds of Air ' (3 vols., 1774–77) and its continuation, ' Experiments and Observations relating to Various Branches of Natural Philosophy ' (3 vols., 1779–86), the six volumes being summarized and systematized in the three volumes of ' Experiments and Observations on Different Kinds of Air and other Branches of Natural Philosophy ' (1790). In these he announced the discovery of oxygen (1775), alkaline air (ammonia) (1774), nitrous oxide (1774–7), vitriolic acid air (sulphur dioxide) (1775), and nitrogen peroxide (1777). His experiments on the oxides of nitrogen are important.

Antoine Laurent Lavoisier (14) was born in Paris in 1743. He was well educated in physics and mathematics as well as in chemistry. He entered the Academy in 1768 and became a *fermier général* in 1780. He carried out many researches of national importance and was a member of the commission for the introduction of the metric system. He had ample means and had a liking for elaborate and expensive apparatus. His work was carefully planned and strictly quantitative in character. In a large number of memoirs (over sixty in the period 1768–87) he developed the new theory that combustion and calcination depend on the union of the combustible or metal with dephlogisticated air, which he called oxygen, this theory being definitely stated in 1777. It should not be forgotten that Hooke in 1665 and Mayow in 1674 had made some striking anticipations of the theory that the atmosphere contains a substance which exists in nitre, although they did not isolate this " nitre air " (oxygen). The Russian scientist Lomonossov in 1744–7 had also made experiments on calcination, showing that there was no increase in weight with sealed retorts.

Lavoisier, in collaboration with other French chemists, introduced a new nomenclature (1787), and his text-book, ' Traité Élémentaire de Chimie ' (1789) systematized the new theory and made it well-known. It was rapidly adopted. Lavoisier's work on the analysis of organic substances by combustion in oxygen laid the foundations of quantitative organic analysis, and his work on respiration, showing that oxygen is absorbed and carbon dioxide evolved, the " combustion " of carbonaceous material being the source of animal heat, was fundamental. Lavoisier's death by the guillotine in 1794 ended a brilliant career, still full of promise.

Lavoisier's general method was to show that when substances burn in air or oxygen, the weight of the product is equal to the sum of the weights of the combustible and the oxygen gas used up. In the calcination of metals, the weight of the calx is equal to the weight of the metal and of the oxygen absorbed. In his most famous experiment, he showed that when mercury is heated in a confined volume of air it slowly forms red calx (oxide) of mercury, the increase in weight being equal to the loss in weight of the air. On heating the calx, pure oxygen, the weight of which was equal to the loss in weight of the air, was evolved and metallic mercury formed.

Lavoisier defined an " element or principle " as " the last point which analysis is capable of reaching," and he emphasized that " they act with regard to us as simple substances, and we ought never to suppose them compounded until experiment and observation have proved them to be so." On this basis he drew up a fairly extensive table of chemical elements, some of which were later proved to be compounds.

Priestley's work (1772–9) on the part which green plants play in the purification of the atmosphere vitiated by respiration (later shown by others to be due to the conversion of carbon dioxide into oxygen under the influence of light), and that on respiration (1776), were obscured by his use of the phlogiston theory.

Priestley obtained oxygen on 1st August, 1774, at Calne. He was heating " a great variety of substances " in glass tubes over mercury by a burning glass, and observed that an " air " was evolved from red precipitate of mercury (oxide of mercury) which supported combustion much better than common air. He had obtained this gas in 1772 by heating nitre, but had not followed up the observation. The discovery seems to have been accidental, since he says that " so far from having formed any hypothesis that led to the discoveries I made . . . they would have appeared very improbable to me had I been told of them." In March, 1775, he knew that the " dephlogisticated air " (*i. e.*, common air *minus* phlogiston) was different from nitrous oxide, which also supported combustion better than common air.

The gas left after combustions had occurred in common air, and any fixed air (carbon dioxide) formed had been extracted by alkali, was called phlogisticated air by Priestley. He regarded it as common air loaded with phlogiston. This was nitrogen. Nitrogen was discovered about 1772 by Scheele, who had recognized it as a distinct gas present in air ; by Cavendish, who communicated his result to Priestley (as the latter says in his paper of 1772) ; and by Daniel Rutherford (' Dissertatio inauguralis de Aere fixo dicto, aut mephitico,' Edinburgh, 12th September, 1772).

In October, 1774, Priestley told Lavoisier in Paris of his discovery of the " air " obtained by heating red precipitate, saying that it supported combustion much better than common air. By 1776 Lavoisier had found that common air is formed by mixing dephlogisticated air and phlogisticated air. In 1777 he was able to state that air is composed of two different elastic fluids, one respirable and capable of combining with metals to form calces, and the other (which he then called a *mofette*, but afterwards named azote) incapable of sustaining respiration and combustion. This was published in 1778, and in 1777 exactly the same statement had appeared in the belatedly published book of Scheele.

The phlogiston theory gave an explanation of the fire and light disengaged in combustion processes. These were due to the escape of phlogiston, the principle of fire, which streamed out of the burning body in a rapid whirling or vortex motion. Lavoisier, in showing that combustion involves a gain of oxygen and not a loss of phlogiston by the body, had told only half the story; the production of fire had still to be explained. He did not distinguish sharply between heat and light, and in place of the weightless phlogiston (sometimes regarded as having a negative weight) he introduced the weightless material of fire, which he called *caloric*. Gases are compounds of material bases and caloric, just as, according to Black, steam is a compound of liquid water and latent heat. Oxygen gas is a compound of oxygen base and caloric; in combustion the oxygen base combines with the combustible, such as phosphorus, sulphur, or carbon, and caloric is set free in the form of fire, or heat and light. The phlogiston theory regarded the combustible body as the source of the fire, Lavoisier considered that this came from the oxygen gas.

The caloric theory of heat was very generally adopted. In Lavoisier's work with Laplace [15] on the ice calorimeter the alternative theory that heat was some form of molecular motion was proposed, but almost certainly by Laplace, who was then engaged in adapting the principles of dynamics to an explanation of physical changes, and hence favoured a dynamical theory of heat (which goes back to Francis Bacon). Lavoisier was more attracted by the material, or chemical, theory of heat, which regarded it as an imponderable element, and at that time it certainly seemed more plausible. He develops this idea quite dogmatically, and the caloric theory was generally accepted. One of Davy's first researches [16] was an attempt to refute Lavoisier's caloric theory experimentally; his later work on the alkalis and chlorine provided further corrections to Lavoisier's views.

In its later phases the phlogiston theory took many curious forms [17]. The necessity of explaining the heat and light in combustion remained an important part of chemical theory. The statement of the various forms of the phlogiston theory is a highly technical part of the history of chemistry. One phase of the theory is concerned with the supposed levity of phlogiston. In this, a distinction was drawn between relative levity, phlogiston being regarded as lighter than air (Guyton de Morveau, 1772), and absolute levity, phlogiston being regarded as absolutely light, and hence decreasing the weight of a body with which it unites. The absolute levity of phlogiston had been entertained by Stahl (1747), and was held for some time by Gren (1786). Its consequences engaged the attention of mathematical physicists (J. T. Mayer, 1790; C. F. Hindenburg, 1790), whose criticisms forced Gren to abandon the theory.

A third phase of the phlogiston theory was held after the general adoption of Lavoisier's theory, since the explanation of the source of fire given by Lavoisier was not always felt to be satisfactory. This form of the phlogiston theory seems to have been first proposed by Gadolin (1788); it was developed by Richter (1793) and by Gren (1794). It started from Macquer's theory (1778) that phlogiston "is nothing but the proper substance of light, immediately or mediately fixed in a large number of compounds," *i. e.*, the matter of light, now distinguished from the matter of heat or caloric. In the development of this theory many took part, but its most complete form is presented by Richter, the less known theories of Crawford (1779–88), Elliot (1780–82), Hopson (1781), Lubbock (1784), and Scott (1786) differing from his only in points of detail.

Lavoisier supposed that combustion is an example of simple affinity, the combustible taking the oxygen-base from oxygen gas and letting the heat-matter (caloric) of the oxygen gas escape. In this case, light must be considered to be a modification of heat-matter, which had not been proved, or else the source of the light must be regarded as different from that of the heat; yet both the cause of the light and the matter of heat are, on this theory, located in the oxygen gas. It is more likely that combustion depends on double affinity, in which case something is present in the combustible body which is called combustible matter or phlogiston, the combustible being a compound of an unknown substrate and phlogiston. Light is a compound of phlogiston and the matter of heat. Every combustion in atmospheric air or oxygen is an action of double affinity; the oxygen-base combines with the combustible substrate, and the matter of heat of the oxygen gas combines with the phlogiston of the combustible to form heat and light, the relative proportions of which depend on the amounts of matter of heat and phlogiston available. In the decomposition of oxide of mercury by heat, the substrate which is combined with oxygen takes up phlogiston from the fire, in which phlogiston is combined with the matter of heat, and at the same time the oxygen-base combines with the matter of heat to form oxygen gas. Thus: combustible=base+phlogiston; oxygen gas=oxygen base+ matter of heat; fire=phlogiston+matter of heat.

Richter gave a searching criticism of Lavoisier's theory, and answered several objections to the phlogiston theory. It was not until the theory of energy was fully recognized, in the middle of the nineteenth century, that chemists could feel that the problem of the production of fire in combustion no longer concerned them directly. Until then, chemistry included the imponderable elements heat, light, electricity, and magnetism, the disentanglement of which into energy and other entities is still proceeding.

Lavoisier's new theory of combustion left an important chemical change unexplained. A metal such as zinc dissolves in dilute sulphuric acid to form sulphate of zinc and an inflammable gas, called inflammable air, is evolved. Sulphate of zinc is formed when calx of zinc (oxide of zinc) dissolves in dilute sulphuric acid, but no inflammable air is evolved. On the phlogiston theory these two reactions were easily explained.

Inflammable air had been assumed by Cavendish (1766), Kirwan (1781), and Priestley (1782), to be identical with phlogiston, and zinc is a compound of calx of zinc and phlogiston, hence :

$$\text{I. } \underbrace{(\text{calx of zinc} + \phi)}_{\text{zinc}} + \text{acid} = \underbrace{(\text{calx of zinc} + \text{acid})}_{\text{zinc salt}} + \underbrace{\phi}_{\text{inflammable air}}$$

$$\text{II. } \text{calx of zinc} + \text{acid} = \underbrace{(\text{calx of zinc} + \text{acid})}_{\text{zinc salt}}$$

Lavoisier gave no explanation of this puzzle, the solution of which awaited the discovery of the composition of water.

In 1781 Warltire, an itinerant lecturer on science, wrote to Priestley saying that he had exploded a mixture of dephlogisticated air (oxygen) and inflammable air (hydrogen) in a copper vessel by an electric spark, and had found a small loss in weight ; also that he had exploded the mixed gas in a glass vessel (as Priestley had done in 1775) and observed, as he says Priestley had done, that the glass, though clean and dry before, became bedewed with moisture. Priestley in reporting this says it confirmed an opinion long entertained by Warltire, that common air deposits its moisture when it is phlogisticated. This " random experiment," as Priestley calls it, was communicated to Cavendish [18], who began work on the subject in 1781. He soon showed that there was no loss in weight, and then, by exploding a mixture of dephlogisticated air and inflammable air in a closed glass vessel by a spark, he obtained water, the weight of which was equal to the weight of the two gases taken. The formation of an acid in the water was a discovery which protracted Cavendish's experiments until he had shown that it was nitric acid, formed from a little nitrogen contained as an impurity in the gases, and his paper was not read until January, 1784. Cavendish stated his results in a peculiar form. He says :

> " I think we must allow that dephlogisticated air is in reality nothing but dephlogisticated water, or water deprived of its phlogiston ; or, in other words, that water consists of dephlogisticated air united to phlogiston ; and that inflammable air is either pure phlogiston, as Dr. Priestley and Mr. Kirwan suppose, or else water united to phlogiston ; since, according to this supposition, these substances united together to form pure water . . . there seems the utmost reason to think that dephlogisticated air is only water deprived of its phlogiston, and that inflammable air, as was said before, is either phlogisticated water, or else pure phlogiston ; but in all probability the former."

In a footnote he explains the reason for his preference. Whereas common or dephlogisticated air will combine at the ordinary temperature with nitrous air (nitric oxide), taking the phlogiston from it (*i. e.*, oxidizing it) and forming red fumes of " phlogisticated nitrous acid " (nitrogen peroxide), dephlogisticated air will not act upon inflammable air below a red heat. Hence, inflammable air can hardly be pure phlogiston, which would be expected to unite at once with dephlogisticated air.

Cavendish thus represented his results as follows :

$$\text{inflammable air} = \text{water} + \phi,$$
$$\text{dephlogisticated air} = \text{water} - \phi,$$
$$\text{inflammable air} + \text{dephlogisticated air} = 2 \text{ water}.$$

He seems to have regarded water as pre-existing in both gases, and to be deposited as the result of a redistribution of phlogiston. In the paper, however, Cavendish says his experiments could also be explained by the assumption that :

> " . . . water consists of inflammable air united to dephlogisticated air ; . . . and indeed, as adding dephlogisticated air to a body comes to the same thing as depriving it of its phlogiston and adding water to it, and as there are, perhaps, no bodies entirely destitute of water, and as I know of no way by which phlogiston can be transferred from one body to another, without leaving it uncertain whether water is not at the same time transferred, it will be very difficult to determine by experiment which of these opinions is the truest ; but as the commonly received principle of phlogiston explains all phaenomena, at least as well as Mr. Lavoisier's, I have adhered to that."

Lavoisier at first thought the product of combustion of inflammable air in oxygen should be an acid, which he tried in vain to find. In June, 1783, Blagden (who had been Cavendish's assistant) visited Lavoisier in Paris and told him that Cavendish had shown that on exploding inflammable air and dephlogisticated air, water equal in weight to the two airs was the only product. A hurried experiment was made by Lavoisier and some other French scientists, and on the next day, or within a few days, an account of it was sent to the Academy. A summary was published in December, 1783, in Rozier's 'Observations sur la Physique,' in which Cavendish's results are most inadequately mentioned. The paper published in the memoirs of the Academy in 1784 contains the remarkable statement that Blagden :

> 'told us that M. Cavendish had already tried, in London, to burn inflammable air in closed vessels, and that he had obtained a very sensible amount of water.'

Lavoisier did not show that the weight of water formed was equal to the weights of the two gases, merely remarking that, since :

> 'it is no less true in physics as in geometry that the whole is equal to its parts . . . we think we are right in concluding that the weight of this water was equal to that of the two airs which served to form it.'

However true this might be in geometry and physics, chemists could feel that it was no adequate substitute for an experiment in which the result had been demonstrated, and such an experiment had been made by Cavendish. Blagden afterwards [(19)] pointed out that Lavoisier should have said in his memoir that he had been told of Cavendish's experiments and that Cavendish had obtained, not merely a very sensible quantity of water, but a quantity of water equal in weight to the sum of the weights of the two gases, adding that Lavoisier had discovered nothing beyond what had been pointed out to him as having previously been done in England.

Blagden, a close friend of many French chemists, lived much in Paris, and must have been very sure of his ground before publishing this statement. In a later report by Lavoisier, Brisson, Meusnier, and Laplace, it is, in fact, stated that :

> 'Cavendish seems first to have noticed (*remarqué*) that the water formed in this combustion is the result of the combination of the two aerial fluids, and that its weight is equal to theirs. Many experiments made on a large scale by Lavoisier, la Place, Monge, Meusnier, and Lefebvre de Gineau, have confirmed this important discovery.'

In his first publication in December, 1783, and in his memoir of 1784, Lavoisier gave, for the first time, a clear statement of the composition of water. Although it is necessary, he says to prove that the matter of heat and light which escapes from the vessel when the two gases combine has no weight, and this was engaging his attention, it could hardly be doubted that water :

> 'previously regarded as an element, is not a compound body . . . formed by the union of the oxygen principle with the inflammable aqueous principle.'

With the discovery of the composition of water, the source of the inflammable air formed when a metal dissolves in a dilute acid is clear. The metal withdraws oxygen from the water to form the oxide, which dissolves in the acid to form a salt, and the hydrogen of the water is set free as the inflammable air :

$$\underbrace{(\text{hydrogen}+\text{oxygen})}_{\text{water}}+\text{zinc}=\underbrace{(\text{zinc}+\text{oxygen})}_{\text{oxide of zinc}}+\text{hydrogen}$$

$$\text{oxide of zinc}+\text{acid}=\text{zinc salt.}$$

It was truly said by Davy that the explanation of the phenomena of combustion, calcination, and (in part) of respiration by Lavoisier was only part of the advance in chemical knowledge made in the eighteenth century. Another, and no less important part, was concerned with the quantitative laws of chemical combination, and the investigation of the laws of affinity.

It is a mistake to suppose that Lavoisier introduced the use of the balance and quantitative method into chemistry. Quite apart from the quantitative experiments of Black, there had been some good quantitative work in the seventeenth century, and in the eighteenth century there was careful quantitative work by Cavendish and Bergman before Lavoisier, and by Wenzel, Richter and Kirwan after he had begun his experiments. The law of conservation of mass, formally stated by Lavoisier (1789), had been assumed in quantative analyses for a long time, and no one would have spent time and trouble in making quantitative analyses of salts unless these were supposed to have definite compositions. The law of fixed proportions was formally stated by Proust in 1797. The law of equivalent proportions was first stated by Richter [(21)] in 1791, although Cavendish in 1766 and 1788 had recognized particular cases of it, and had used the name equivalent. Owing to a slip by Berzelius, the law of equivalents was attributed to Wenzel (1777), who had actually denied its validity.

Jeremias Benjamin Richter was born in Silesia in 1762. He studied under Kant at Königsberg, and graduated M.D. in 1789 with a dissertation 'De Usu Matheseos in Chymia.' Before this he was in an engineer corps in the army, and had been studying chemistry. He failed to obtain an academic position, and in 1798 became chemist in the Berlin porcelain factory. He carried out, mostly at night, his own experimental and literary work, and made hydrometers for sale. His health was poor, and during his short life he received little encouragement; even after his death his discoveries were mostly credited to others. Besides his work on combining proportions, he made important investigations on uranium and other subjects.

Richter was permeated with the idea, which he may have got from Kant, that chemistry is a branch of applied mathematics. His style is generally tedious, but now and again statements of great beauty and power stand out from the dull background of mathematical irrelevance. Richter was a man of great independence and originality. Arithmetical errors in his publications drew upon him very unfavourable and wounding comments, the great importance of his generalizations being quite overlooked.

In 1791 Richter found that solutions of calcium acetate and potassium tartrate remain neutral on mixing, although calcium tartrate is precipitated and potassium acetate remains in solution. From this and similar experiments he drew the conclusion that:

> 'there must be a definite ratio between the masses of each neutral compound and that the terms of ratio are of such a character that they can be determined from the masses of the neutral compounds.'

In 1792 he expressed the same idea as follows:

> 'when two neutral solutions are mixed, and a decomposition follows, the new resulting products are almost without exception neutral also . . . the elements must therefore have among themselves a certain fixed proportion of mass.'

The composition of the products can be calculated from those of the reacting substances:

> 'This rule is the true touchstone of the experiments instituted with regard to the ratios of neutrality; for if the proportions empirically found are not of the kind that is required by the law of decomposition by double affinity, where the decomposition actually taking place is accompanied by unchanged neutrality, they are to be rejected without further examination as incorrect.'

Richter drew up tables of the combining proportions (equivalents) of acids and bases, and attempted to find non-existent mathematical regularities in them. For some reason he failed to see that all these numbers can be combined in a single table, between the members of which ratios of equivalence hold good. This important step was taken by Ernest Gottfried Fischer (1754–1831) of Berlin, in his German translation of Berthollet's 'Recherches sur les Lois de l'Affinité' (Berlin, 1802), in which he gave an account of Richter's little known researches and drew up a table of equivalents of acids and bases, taking 1000 parts of sulphuric acid as standard. Fischer's table, and part of his note, were reproduced in a French translation in the first volume of Berthollet's 'Statique Chemique' (1803, I, 134), which made Richter's work known, his own writings being extremely scarce even at the beginning of the nineteenth century. The only nearly complete set of them in England, found, with much trouble, by Angus Smith, is in the Manchester University Library.

Another set of quantitative experiments on combining proportions was made by Richard Kirwan [22], born in Galway in 1733. Educated in law and for a time practising as a barrister in London, Kirwan turned to science, in 1799 becoming President of the Royal Irish Academy in Dublin. He was a man of great brilliance, versatility and originality. He was at first a defender of the phlogiston theory, his 'Essay on Phogiston and the Constitution of Acids' (London, 1787) being translated into French (1788) with refutations of the various parts by Lavoisier, Berthollet, Fourcroy, de Morveau, and Monge. In 1791 Kirwan announced his conversion to the antiphlogistic theory.

Kirwan also drew up the first table of specific heats (1780), and wrote 'Elements of Mineralogy' (1784; second edition 1794–6), containing a scale of hardness based on Werner's, and very like that afterwards called Mohs' scale.

In a series of papers [23] Kirwan published measurements of the specific gravities of salt solutions and the compositions of salts. These really include tables of equivalent weights, although he overlooked this fact. In his later papers he refers to Richter. He also determined the densities of gases.

The explanation of the laws of chemical combination, the law of constant proportions, the law of equivalents and the law of multiple proportions (implied in the statements of William Higgins in 1789) was first adequately given by Dalton early in the nineteenth century, but some interesting approximations to the chemical atomic theory were made by Bryan Higgins, born in Sligo about 1737, who conducted a school of practical chemistry in Greek Street, Soho (which Priestley attended), and his nephew William Higgins, born in Sligo about 1768, who was assistant to Beddoes at Oxford and later professor in Dublin. Bryan Higgins [24] believed that the particles of elements attract one another, and all of them except earth and water are also endowed with repulsive forces. The repulsion is due to fire, which is a material elastic fluid pervading all bodies. In gases, the particles stand at some distance from one another, and combination (or "saturation") cannot occur until the atmospheres of fire round the particles are broken. By means of diagrams he shows how ammonia and hydrogen chloride (which he calls "alkaline fluid" and "acid fluid," respectively) unite particle with particle to form solid sal ammoniac; this combination is in "determinate proportion," and the "molecules" of the compound contain one atom of each "element," because atoms of like nature repel each other. This was an extension of Newton's theory of gases in which it is combined with Newton's theory of chemical combination.

William Higgins was one of the first British chemists to accept Lavoisier's views, his book 'A Comparative View of the Phlogistic and Antiphlogistic Theories' (1789, second edition 1791) being a reply to the 'Essay on Phlogiston' of his friend Kirwan. He said [25]:

> 'For a considerable time I stood alone in England, where I then resided, being the first who adopted the antiphlogistic doctrine, and the only man who had expressly written in favour of it in the English language.'

This is perhaps literally correct, since Black, who had accepted Lavoisier's theory before 1784, lived in Scotland, and the spirited attack on phlogiston and defence of the antiphlogistic theory written in 1784 by John Lubbock ('Dissertatio de principio sorbile,' Edinburgh, 1784) was in Latin.

William Higgins recognized that the existence of several oxides of nitrogen proved that the assumption that combination (or "saturation") occurs only by the union of one particle of each element is too restricted. He recognized a decrease in stability as more and more particles of one element are added to a particle of another, and explained this by assuming that the force of attraction between a particle A is equally shared between successive particles of B until, in the end, the attractive force becomes too small to hold the particles together. This force is quite different from gravitational force, and cannot be said to be satisfactory. The law of multiple proportions follows at once from the description and diagrams. In all but one case, he assumed that the particles or atoms (which he calls "molecules") of different elements are equal in weight; the equivalents of common elements (except hydrogen) are, in fact, not very different (carbon 6, nitrogen 7, oxygen 8, sulphur 8). Thus he did not anticipate Dalton's clear statement that particles of different elements have different weights, and he says that equal volumes of gases contain unequal numbers of particles (as Dalton also believed).

We have now reached the end of the eighteenth century, and a summary may be given of some of the important discoveries in chemistry made during that century and the position of the science at its close.

One of the most important contributions was the clear recognition of several kinds of gases, different in physical and chemical properties. Hydrogen, oxygen, nitrogen, chlorine, the oxides of nitrogen, carbon dioxide, carbon monoxide, ammonia, sulphur dioxide, and silicon fluoride were all known. In their discovery, Scheele and Priestley had played the most prominent part. Closely connected with the discovery of oxygen was the development by Lavoisier of a new theory of combustion, which replaced the phlogiston theory held earlier in the century.

Lavoisier's theory also required a revision of the list of elements, since many substances regarded as elements by the phlogiston theory were shown to be compounds, and *vice versa.* This made it necessary to introduce a new nomenclature in chemistry, which was done by Lavoisier and his associates. The chemical nature of respiration, and the changes taking place in the atmosphere in the growth of green plants in light, were also fairly clearly stated in the later years of the century.

Elements discovered in the eighteenth century are chlorine (Scheele, 1774), chromium (Vauquelin, 1797), cobalt (Brandt, 1735), manganese (Gahn, 1774), molybdenum (Hjelm, 1790), nickel (Cronstedt, 1751), nitrogen (Scheele, Cavendish, Priestley, Rutherford, 1772), oxygen (Scheele, 1772; Priestley, 1774–5), platinum (Brownrigg, 1750), strontium (as compounds, Crawford, 1790; the element by Davy, 1808), tellurium (Müller von Reichenstein, 1782), titanium (Gregor, 1789), tungsten (brothers d'Elhujar, 1783), uranium (Klaproth, 1789), zirconium (as compounds, Klaproth, 1789, the element by Berzelius, 1824).

The mathematical-quantitative method introduced into astronomy and physics by Newton, and applied by him and his followers in chemistry, had proved unfruitful in several lines of research in which it was too closely followed, and the chemical method, which had arisen from alchemy but was freed from mysticism and made rational by Van Helmont and Boyle, had led to very important discoveries in the seventeenth century. In the eighteenth century it proved very fruitful in the hands of such experimenters as Scheele and Priestley, who either made good use of qualitative and incorrect theories, or (in Priestley's case) had very little interest in theory of any kind. Lavoisier, a master of quantitative method, made no discoveries of chemical substances, and his main contribution to the science consisted in drawing correct conclusions from the discoveries of others, and reinforcing them by his own experiments.

Although some quantitative experiments on the composition of chemical compounds had been made in the seventeenth century, this branch of chemistry was greatly extended in the eighteenth century by the experiments of Cavendish, Wenzel (who failed to draw correct conclusions from his results), Bergman, Richter and Kirwan. In this way the laws of fixed, equivalent, and multiple proportions were perceived (the last by William Higgins). Richter's tables of equivalents were combined later into one by Fischer (1802). Bryan and William Higgins speculated on the combination of particles in a way which could have led them to the chemical atomic theory, but did not.

Although the phlogiston theory was abandoned after Lavoisier's work, it left the explanation of the evolution of heat and light in chemical changes still undecided, and several modifications of the theory were proposed in which the existence of phlogiston was assumed alongside that of oxygen. This problem still remained until the middle of the nineteenth century, when the theory of energy was accepted.

The eighteenth century was perhaps the most important one in the development of chemistry. It saw the introduction of a correct view of the chemical elements, a true explanation of the most important chemical changes, and the discovery of a large number of important substances. With this foundation, the science could develop during the nineteenth century on a plan laid down by the atomic theory of Dalton. The eighteenth century was the heroic age of chemistry.

References.

(1) Boyle, 'Works,' edit. Birch, i. Life, p. 71 (1744).

(2) Phil. Trans., xxvi. p. 97 (1708).

(3) 'Praelectiones Chymicæ,' London (1709); 'Chymical Lectures,' London (1712); 'Opera Omnia Medica,' London (1733).

(4) *Mém. Acad. R. Sci.*, p. 202 (1718); p. 20 (1720); Macquer, 'Élémens de Chymie theorique,' pp. 256–73 (1749).

(5) 'De attractionibus electivis,' *Nova Acta Upsal.*, ii. pp. 159–248 (1775); 'Opuscula,' Uppsala, iii. p. 291 f (1783).

(6) J. R. Partington, 'The Origins of the Atomic Theory,' Annals of Science, iv. p. 245 (1939).

(7) J. R. Partington, 'John Dalton,' Endeavour, vii. p. 54 (1948).

(8) J. R. Partington, 'Joan Baptista Van Helmont,' Annals of Science, i. p. 359 (1936).

(9) D. McKie, Phil. Mag., xxxiii, p. 847 (1942).

(10) 'Experiments on Factitious Airs,' Phil. Trans., lvi. p. 141 (1766); 'Scientific Papers,' Cambridge, ii. p. 77 (1921).

(11) 'Essays and Observations, Physical and Literary, read before a Society in Edinburgh, and published by them,' Edinburgh, ii. pp. 157–225 (1756); Alembic Club Reprint, No. i.

(12) A. E. Nordenskiöld, 'Carl Wilhelm Scheele. Nachgelassene Briefe und Aufzeichnungen,' Stockholm (1892); 'Collected Papers of Carl Wilhelm Scheele,' edit. L. Dobbin, London (1931); C. W. Oseen, 'Torbern Bergman och Carl Wilhelm Scheele,' Stockholm (1940) (Levnadsteckningar över K. Svenska Vetenskapsakademiens ledamöter, p. 113); *idem.*, 'Carl Wilhelm Scheele. Handschriften 1756–1777, Ausdehnung,' Stockholm, K. Svenska Vetenskapsakademien (1942); A. Fredga, 'Carl Wilhelm Scheele,' Stockholm (1943) (Levnadsteckningar över K. Svenska Vetenskapsakademiens ledamöter, p. 119).

(13) J. R. Partington, Nature, cxxxi. p. 348 (1933); D. McKie, 'Joseph Priestley, Chemist,' Science Progress, xxviii. p. 17 (1933); Sir P. J. Hartog, 'The Newer Views of Priestley and Lavoisier,' Annals of Science, v. p. 1 (1941).

(14) Lavoisier, 'OEuvres,' 6 vols., Paris (1864–93); J. A. Cochrane, 'Lavoisier,' London (1931); D. McKie, 'Antoine Lavoisier,' London (1935); see. ref. 13.

(15) *Mém. Acad. R. Sci.*, p. 355 (1780 [1784]); 'OEuvres,' ii. p. 283.

(16) H. Davy, 'An Essay on Heat, Light, and the Combinations of Light,' 'An Essay on the Generation of Phosoxygen (Oxygen Gas)'; in 'Contributions to Physical and Medical Knowledge, Principally from the West of England,' Bristol, pp. 5, 151 (1799); Davy, 'Works,' ii. p. 1 f (1839).

(17) J. R. Partington and D. McKie, 'Historical Studies on the Phlogiston Theory,' Parts i–iv; Annals of Science, ii. p. 361 (1937); iii. pp. 1, 337 (1938); iv. p. 113 (1939).

(18) 'Experiments on Air,' Phil. Trans., lxxiv. p. 119 (1784); lxxv. p. 372 (1785); 'Scientific Papers,' Cambridge, ii. pp. 161, 187 (1921); J. R. Partington, 'The Composition of Water' ('Classics of Scientific Method'), London (1928).

(19) Crell, *Chemische Annalen*, i. p. 95 (1784).

(20) *Annales de Chimie*, vii. p. 257 (1790).

(21) J. B. Richter, 'Ueber die neuern Gegenstände der Chemie,' Parts i–xi, Breslau and Hirschberg (1791–1802); 'Anfangsgründe der Stöchyometrie oder Messkunst chymischer Elemente,' 3 vols. (in 4), Breslau and Hirschberg (1792–94); R. Angus Smith, Mem. Manchester Lit. and Phil. Soc., xiii. p. 186 (1856).

(22) On Kirwan, see J. Reilly and N. O'Flynn, Isis, xiii. p. 298 (1930); D. Reilly, 'Three Centuries of Irish Chemists,' Cork, p. 8 (1941).

(23) Phil. Trans., lxxi. p. 7 (1781); lxxii. p. 179 (1782); lxxiii. p. 15 (1783); Trans. Roy. Irish Acad., iv. p. 1 (1791); vii. p. 163 (1800); viii. p. 53 (1802).

(24) 'A Syllabus of Chemical and Philosophical Enquiries' [1775]; 'A Philosophical Essay Concerning Light,' London (1776); 'Experiments and Observations relating to Acetous Acid . . . . and other Subjects of Chemical Philosophy,' London (1786); 'Minutes of a Society for Philosophical Experiments and Observations,' London (1795); on Bryan and William Higgins, see A. N. Meldrum, New Ireland Review, xxxii. pp. 275, 350 (1909–10); J. Reilly and D. T. MacSweeney, Sci. Proc. Roy. Dublin Soc., xix. p. 139 (1929); on their corpuscular theories, see J. R. Partington, ref. 6.

(25) 'An Essay on Bleaching,' p. xi. (1799).

MATHEMATICS THROUGH THE EIGHTEENTH CENTURY

By J. F. SCOTT, Ph.D.

A. Developments in Pure Mathematics.

The close of the seventeenth century marks a turning point in the history of mathematics. Within a space of less than fifty years, two weapons of enormous power had been forged ; these were the analytical geometry of Descartes and Fermat and the calculus of Newton and Leibniz. But these achievements, momentous though they were, represent but a fraction of the mathematical contributions of the seventeenth century. A workmanlike system of notation had been evolved, and side by side there had grown up an increased knowledge of the nature of equations. The important theory of probability had been foreshadowed by the investigations of Fermat and Pascal. The foundations of geometry had been critically examined, and a new type of geometry, the projective geometry of Desargues, had made its appearance. In mechanics the fundamental ideas on motion had been enunciated by Galileo and clarified by Newton, and with the formulation of the law of gravitation at the hands of the latter, the subject had begun to assume the appearance of a well-coordinated science. It is hardly too much to say that in regard to its mathematical development, no century starting from so little had achieved so much.

The eighteenth century proved a worthy heir to its mathematical heritage. Mathematical discovery continued to advance even with accelerated pace and in more diverse directions. But of the vast fields now thrown open to the mathematician, none proved so attractive as the new calculus, and for the first few decades mathematicians seemed to have little care beyond improving its technique and widening the scope of its application. The new instrument was wielded with ingenuity and with power and this led to a marked expansion of the mathematician's horizon. Within a few years such offshoots as the calculus of finite differences and the calculus of variations began to appear. Specialized branches as, for example, the study of differential equations, elliptic integrals and the theory of probability were established and prosecuted with vigour. Not the least important aspect of this awakening of mathematical interest was the application of the new methods to problems in mechanics, especially celestial mechanics. Interest in the geometrical methods of Descartes waned somewhat during the latter half of the seventeenth century; the calculus was thought to prove a richer mine, and it was not until the eighteenth century was well established that attention commensurate with their importance began to be directed to them.

In this mathematical renaissance many countries took part. The English school at first appeared vigorous and fruitful, but interest was short-lived and after the middle of the century Englishmen had little share in the developments which were taking place all over Europe. This was largely a result of an excessive reliance upon the geometrical and fluxional methods of Newton. The great German philosopher, Leibniz, had fashioned his calculus with less regard to rigour and strict logic, but he had framed it in a simpler and more workmanlike notation. Moreover, it had found able and zealous disciples in the two Bernoullis, James and John, to whose industry in popularizing the new methods on the continent the calculus owes much. The English school, unwilling to emancipate itself from the trammels of a clumsy notation, lagged behind, and it was not until a century had elapsed that it began to take its rightful place in the front rank. Nevertheless, its contributions were not to be despised. Names such as De Moivre, Cotes, Taylor, Stirling, Maclaurin, Simpson and others remind us of its virility before the decline set in.

But it is to the continent that we must turn for the most spectacular advances. France had taken the lead in the later years of the seventeenth century and she did not relinquish it until well on into the eighteenth. Varignon, L'Hôpital, Parent, Clairaut, D'Alembert were the pioneers; they were succeeded by names no less illustrious in the history of mathematics: Monge, Laplace, Legendre, Poncelet, La Croix, Cauchy. Italy's contributions were far from insignificant, and names such as Ceva, Fagnano, Agnesi, Ruffini bear eloquent testimony to the restless quest for mathematical adventure under southern skies. The greatest mathematician of the century, Lagrange, though of Italian birth, spent his best years outside his native land, and the same may be said of Euler, whose industry and genius have left a permanent impression upon every department of mathematical knowledge. Born at Bâle, Euler spent most of his time at St. Petersburg, and it was thence that he made contributions to almost every branch of mathematics. Bâle may also claim the Bernoullis; of this gifted family it was James and John who gave the greatest impulse to the study of the new methods and made them understandable.

Germany did not show her real strength until the century was well established, even though much of Lagrange's best work was compiled during his residence in Berlin. Gauss, the greatest of a brilliant line, was unknown until nearly the close of the century. It was the particular glory of Gauss that he impressed conceptions of rigour upon the new methods; prior to him even the greatest mathematicians had not paid sufficient attention to the foundations upon which they were erecting such an imposing structure.

A tendency to increased specialization was an inevitable consequence of the rapid widening of the mathematician's domain. We pass, therefore, to a brief account of the main branches which were cultivated during the eighteenth century. It will be convenient to group them under the following headings :—

(1) The extension and systematization of the calculus, and its application to physical problems.

(2) The establishment of the calculus of variations.

(3) The development of the theory of probability.

(4) Progress in algebra and related subjects.

(5) Progress in geometry and trigonometry.

The development of applied mathematics will be treated in a separate section.

1. *Extension and Systematization of the Calculus.*

The calculus as it left the hands of Newton and Leibniz was admired but imperfectly understood, and before any solid development could take place the new methods had to be popularized and made intelligible. In this task the two Bernoullis, James and John, combined indefatigable zeal with uncommon ingenuity. It was probably James who was the first to realize the enormous scope of the new calculus, and his persistent advocacy of the new methods is not the least of his services to mathematics. In a collection of essays published in 1691, he explained the principles of the new analysis in a form which could be understood. He applied the calculus to the solution of a number of physical problems and, by his investigation of isoperimetrical problems (1701), he laid the foundations of the calculus of variations. In 1698 he published an essay on the differential calculus and its application to geometry ; in this he investigated the properties of a number of curves, notably the equiangular spiral. His great work, 'Ars Conjectandi' (1713), published eight years after his death, contains much that is of the highest importance in mathematics. Though its main contribution is to the theory of probability, there is much in it that is still of value in other branches of mathematics. Hardly less influential in developing the new methods was his brother John, whose 'Lectiones Mathematicae de Methodo Integralium' (written 1691, published 1742), was a landmark in the development of the calculus. The quadrature of surfaces, the rectification of curves, the solution of different types of differential equations are all to be found in its pages ; the work, too, is replete with examples of different artifices in integration. He also attacked the problem of the vibration of strings. The exponential calculus has been claimed as one of his discoveries. Like his brother, he studied isoperimetrical curves. It was a pupil of John Bernoulli who provided the first treatise on the subject of the calculus. This was the 'Analyse des Infiniment Petits' (1696) of l'Hôpital, a work which did much to give currency to the notation and methods of the calculus.

The Bernoulli tradition was continued by Daniel, son of John, whose 'Exercitationes quaedam Mathematicae' (Venice, 1724) still further advanced the study of differential equations. His best known work, however, is his 'Hydrodynamica' (1738). In this he reduced all his results to a single principle, the conservation of energy.

In 1734 there appeared 'The Analyst: or, A Discourse addressed to an infidel Mathematician,' by Bishop Berkeley. This contained a devastating, yet able, attack upon the principles of the new methods. To this onslaught, Benjamin Robins replied with 'A Discourse concerning the Nature and Certainty of Sir Isaac Newton's Method of Fluxions' (1735), which was designed to shatter the objections raised by the learned bishop. A more vigorous refutation came from Maclaurin, who published in 1742 his 'Treatise of Fluxions,' the first systematic exposition of the method of fluxions to appear. This contains his well-known series. The work also contains many mathematico-physical problems, the solution of which Maclaurin effected by employing the Newtonian methods. Maclaurin also contributed a number of important papers to the 'Philosophical Transactions.' He was one of the most gifted men of the age, but his adherence to geometrical methods was, perhaps, unfortunate, inasmuch as it encouraged a certain indifference amongst his countrymen to the analytical methods which were being pursued with conspicuous success on the continent. Among other British mathematicians, one of the best known is Brook Taylor, a man of profound inventiveness. Like Maclaurin, he was an enthusiastic admirer of Newton. In his 'Methodus Incrementorum Directa et Inversa' (1715), he laid the foundations of the method of finite differences. He also solved a number of difficult problems in physics, notably on the transverse vibration of strings. Roger Cotes widened the resources of integration by some elegant applications of logarithms and circular arcs ('Logometria: A Treatise on Ratios,' Phil. Trans. xxix.). Stirling and the French born De Moivre extended the scope of the calculus by their use of series. The industrious Thomas Simpson made important contributions in his 'New Treatise of Fluxions' (1737). Finally, John Landen tried to remove the metaphysical difficulties inherent in the conception of fluxions by substituting purely algebraical methods. This was the main purpose of his 'Residual Analysis,' which appeared in 1764. Landen is best remembered by his 'Mathematical Lucubrations' (1755). After Landen, the mathematical genius of the English school slumbered. Writing towards the end of the century, Lalande ('Life of Condorcet') affirms that by 1764 there was not a single first-rate analyst in the whole of England (1).

The Italian mathematicians contributed materially to the development of the new methods. The Riccatis, father and son, and Manfredi, gave solutions of differential equations of the first degree which exhibit no little skill. Fagnano, in his 'Produzioni Matematiche' (1750), showed himself a mathematician of power. His attempts at the rectification of the ellipse and the hyperbola led him to the study of elliptic functions, and his paper ' Teorema da cui, si diduce una nuova misura degli Archi Elittici, Iperbolici et Cyclidali ' (1716) marks the first contribution to this important subject. Fagnano's work proved an inspiration to Euler, who from 1733 occupied the chair of mathematics at St. Petersburg. Despite serious physical handicaps Euler was one of the most prolific writers of any age, and there was scarcely any branch of mathematics which his researches did not enrich. In addition to a vast number of memoirs and other works, Euler produced four monumental treatises in pure mathematics alone :—

(1) ' Introductio in Analysin Infinitorum ' (1748),

(2) ' Institutiones Calculi Differentialis ' (1755),

(3) ' Institutiones Calculi Integralis ' (1768–70),

(4) ' Methodus Inveniendi Lineas Curvas Maximi Minimive Proprietate Gaudentes, sive Solutio Problematis Isoperimetrici ' (1744).

The first of these was designed as an introduction to methods purely analytical. It deals with the theory of equations and the geometry of curves. The idea of a function is developed ; logarithms are defined as exponents and many of the difficulties with regard to the logarithms of negative and imaginary numbers which had embarrassed a number of mathematicians were clarified. The second and third works constitute the most complete and systematic exposition of the calculus that had yet appeared. They contain Euler's investigation of differential equations of the second order and a treatment of the beta and gamma functions which were invented by him. Euler turned to elliptic integrals, but his researches thereon are not distinguished by any marked originality. This was reserved for Legendre. Following his preceptor, John Bernoulli, Euler directed his attention to the problem of isoperimetric curves and this led him to the calculus of variations. His contribution to this important subject formed the subject of the ' Methodus Inveniendi Lineas Curvas,' a work which no doubt Lagrange found very much to his taste. Euler also contributed a work on algebra (' Vollständige Anleitung zur Algebra,' 1770). This contains his proof of the binomial theorem, which, however, is not rigorous. In spite of this defect, Euler's work was of a very high standard and there were few branches of mathematics upon which he did not leave his mark.

As we have indicated, it was the French school which cultivated the new methods with the greatest assiduity in the early years of the century, and although many years were yet to elapse before we meet any name of the first rank, nevertheless there were many whose powerful advocacy of the new methods brought them into prominence. A contemporary of L'Hôpital, Pierre Varignon, assisted materially in this direction ; so did Pierre Rémond de Montmort, who contributed to the theory of finite differences, and later to the theory of probability. These were followed by Clairaut, who showed unusual skill in solving differential equations. Clairaut turned aside from analytical to geometrical methods in his 'Théorie de la Figure de la Terre' (1743), nevertheless he continued to publish important mathematical papers dealing with maxima and minima and the calculus. Contemporary with Clairaut was D'Alembert, whose researches into mechanics and astronomy are his chief title to fame. In one of his later works he investigated the problem of vibrating strings, and was led to the differential equation of the form

$$\frac{\partial^2 u}{\partial t^2}=\frac{\partial^2 u}{\partial x^2},$$

the solution of which he gave in a memoir to the Berlin Academy (1747) as

$$u=\phi(x+t)+\psi(x-t),$$

ϕ and ψ being arbitrary functions. D'Alembert's main contributions however, are to applied mathematics and will be mentioned later.

In 1736 was born Joseph Louis Lagrange at Turin. By the time he was twenty he had established himself as one of the foremost mathematicians of even that illustrious age. Although Lagrange showed himself a master of almost every branch of analysis, it was a general solution of certain isoperimetrical problems which he communicated to Euler that first brought him into prominence. In the course of his demonstration, Lagrange developed the calculus of variations. This was followed by a series of memoirs in which he covered a wide range of subjects including partial differential equations, the theory of numbers, elliptic integrals. He also applied his calculus to the solution of a number of physical and astronomical problems, notably that of the moon's libration. Most of these contributions are to be found in the 'Miscellanea Taurinensia.' Lagrange deduced formulæ from purely analytical considerations and he endeavoured to erect a science on strict mathematical principles entirely divorced from ideas of infinitesimals or of vanishing ratios. This was the subject of his 'Théorie des Fonctions Analytiques' (1797, 2nd edn., 1813). Not the least important of his services to the rapidly expanding subject of mathematics was the inspiration he gave to his successors. His 'Leçons sur le Calcul des Fonctions' (1801, 1804, 1806), by the emphasis it placed upon rigorous methods, undoubtedly guided Cauchy in his 'Cours d'Analyse' (1821). But besides rigour, Lagrange impressed generality upon the subject. He made a careful study of the contributions of his predecessors and showed that all their artifices could be reduced to a uniform procedure. Before Lagrange, individual problems had been solved by the employment of a particular device. Each solution was something of a *tour de force*, and the discovery of the appropriate technique was usually regarded as no mean achievement. With rare insight Lagrange systematized the procedure and substituted for a series of attacks upon isolated problems a unified method of approach. His treatment of partial differential equations led to important solutions during the next fifty years of complicated problems in physics. In his monumental treatise, 'Mécanique Analytique'

(Paris, 1788), Lagrange established the principle of virtual work; from this, with the aid of the calculus of variations, he deduced all the theorems of mechanics and hydrodynamics.

Hardly less important in the development of eighteenth-century mathematics was Pierre Simon Laplace. His contributions to mathematics and mechanics, especially celestial mechanics, show him to have been a man of exceptional gifts. His 'Exposition du Système du Monde' appeared in 1796. In this work Laplace applied the calculus to astronomical problems with consummate skill. This was followed by the 'Mécanique Céleste' (1799–1825), which is usually considered his masterpiece. In addition, Laplace contributed many important memoirs on various branches of higher mathematics, such as the theory of equations, determinants, differential equations, and the solution of complicated physical problems. Much of his most enduring work (*e. g.*, 'Théorie Analytique des Probabilités, 1812') lies outside the period under review. The same is true of the great German mathematician, Karl Friedrich Gauss. But before the end of the century Gauss had already given evidence of his exceptional mathematical power by his 'Disquisitiones Arithmeticae,' much of which had already been made public in the decade preceding its appearance in 1801. Gauss's work proved an inspiration to his successors. As a result of his labours to establish the subject upon sound foundations, mathematics left his hands a very different instrument from what it was when he first directed his attention to it.

Before the close of the century there appeared the 'Traité Élémentaire du Calcul Différentiel et du Calcul Intégral' (1797 and 1819), of La Croix, a man who attained a position of repute through perseverance rather than unusual brilliance. He has no outstanding discovery to his name, nevertheless his works were deservedly popular. They were translated into English and in the hands of the Cambridge mathematical school they were instrumental in introducing the continental notation and methods into this country.

2. *The Establishment of the Calculus of Variations.*

The calculus of variations has proved one of the most fruitful outgrowths of the calculus. Though its development was stimulated by the rapid growth of mechanics during the eighteenth century, its origin is much older and can be traced in the investigation of the curve of quickest descent and the determination of the form of a hanging chain, problems which had exercised the ingenuity of Galileo and his contemporaries. The problem reappears in Leibniz in his method of differentiating a curve whose equation itself is presumed to undergo the minutest alteration.

Euler, who may be regarded as the creator of the subject, defines its scope thus: "The calculus of variations is the method of finding the variation undergone by an expression involving any number of variables, when the values of some or all of these are varied" (2). Thus the problem is reduced to making a given integral a maximum subject to a certain restriction.

As an illustration of the nature of the problem involved we may quote the one which brought the subject into prominence during the seventeenth and eighteenth centuries : "There being given two points, not in the same vertical plane, it is required to find the curve along which a body will slide from one to the other in the least possible time." This is the so-called *brachistochronic path*. Mathematically, the problem is similar to the familiar isoperimetrical problem, namely, the finding of the curve of given length which will enclose the maximum area.

In the development of the theory during the eighteenth century we recognize four stages, and these are associated with the names of the Bernoullis, James and John, Euler, Lagrange, and finally Legendre and his successors.

The foundations of a general theory were laid by James Bernoulli when he solved the problem of finding the curve of quickest descent, which had been proposed by his brother John in 1697. The latter had proposed six questions of considerable difficulty, *e. g.*, "of all the semi-ellipses described on a given horizontal axis, to determine the one which will be traversed by a body rolling along its concave surface in the shortest possible time." These appeared in the 'Journal des Sçavans' (Dec. 2nd, 1697). James' solution is to be found in the 'Acta Eruditorum' for June, 1700 (Jacobi Bernoulli : 'Solutio Propria Problematis Isoperimetrici'). This was followed in May, 1701, by another, Jacobi Bernoulli: 'Analysis Magni Problematis Isoperimetrici.' John had already discovered an elegant solution of the brachistochrone problem ; his brother's solution, however, covered a much wider range and was in fact a masterpiece of clarity and skill. The generality of his methods was recognized by Euler, whose attention had for some time been directed to the problem. It was by solving these that he was led to the calculus of variations. In his treatise of 1744 ('Methodus Inveniendi,' etc.), he displayed all his customary skill. But it is Lagrange who must be regarded as having placed the subject of firm foundations. In 1755 he sent Euler a solution of a certain isoperimetrical problem which had been engaging the latter's attention. Euler at once recognized the superiority of Lagrange's method and he did not hesitate to proclaim as much in a memoir, "Elementa Calculi Variationum" (Novi Commentarii Acad. S. Petropolitanae, Tom x). Within five years Lagrange had acquired complete mastery over the new method. It was he who greatly extended the subject and introduced the notation for variation whence the theory has taken its name. His method was strictly analytical and he employed differential equations freely. His methods and results are to be found in the second volume of the 'Miscellanea Taurinensia' (1760–2), and in his 'Théorie des Fonctions Analytiques' as well as his 'Leçons sur le Calcul des Fonctions.'

Euler in his treatise of 1744 had found the necessary condition for the case of one variable. Eleven years later Lagrange found it for two. But neither of these established criteria for distinguishing between maxima and minima. This was the contribution of Legendre, who in

1786 studied for the first time what is called the second variation of the quantity to be minimized. This was the purpose of his "Mémoire sur la Manière de distinguer les Maxima et les Minima" (Hist. de l'Acad. Royale des Sciences, 1786, pp. 7–37). Legendre's work was not free from blemish, nevertheless it inspired others, notably Gauss, Poisson, Cauchy, Jacobi, to remove the obscurities and to develop the subject further. After Jacobi, no further progress was made until Weierstrass directed his attention to it half a century later.

In mechanics the calculus of variations has played a significant part. Lagrange showed its power when he was able to establish by its aid the whole of mechanics on a single principle. It has also inspired later writers and has given rise to a new type of problem, namely that of Least Action, foreshadowed by Fermat in his optical researches and developed by Hamilton.

[See Todhunter: 'History of the Calculus of Variations.' London, 1861.]

3. *The Development of the Theory of Probability.*

It is customary to regard Fermat and Pascal as joint founders of the mathematical theory of probability. But attempts to forecast the results of games of chance are much older. Many mathematicians of the fifteenth and sixteenth centuries—Pacioli, Cardan, Tartaglia, Galileo—to name but few, had applied their skill to the problems involved. Nevertheless, it would be too much to say that the origin of the mathematical theory is discernible before the publication of the celebrated problem submitted by the Chevalier de Méré to Pascal and sent by him to Fermat:

"Two players of equal skill leave the table before the completion of the game. Given the total stakes and the score of each player the problem is to discover how the stakes should be divided."

The problem opened up a new world to Fermat and to Pascal, each of whom gave a solution. For his solution Fermat employed the method depending upon combinatorial analysis, *i. e.*, the finding of the number of ways in which a given operation can be performed, or in which an event, completely specified, can take place. Pascal, who arrived at the same solution as Fermat, relied upon his "arithmetical triangle." Pascal's methods were followed by Huygens, from whom the next contribution came; this was the 'De Ratiociniis in Ludo Aleae,' which Schooten inserted at the end of his 'Exercitationes Mathematicae' in 1658. In spite of the widespread interest which the subject had aroused, nothing further appeared for half a century, when Pierre Rémond de Montmort, a pupil of Malebranche, published a work on chances which contained some novelties ('Essai d'Analyse sur les Jeux de Hasard,' 1708). Meanwhile, James Bernoulli had made some trivial observations upon the subject in the 'Journal des Sçavans' (1685) and the 'Acta Eruditorum' (1690). These were elaborated and developed in the 'Ars Conjectandi' (1713), published posthumously. This may be regarded as the first serious contribution to the theory of the subject. The treatise is divided into four parts :—(i) A statement of the principles of the theory, together with a reprint of the problems of Huygens, (ii) The theory of the doctrine of permutations and combinations, (iii) The solution of a large number of problems relating to games of chance, (iv) Application to moral and economic questions. Despite the comparative novelty of the subject, Bernoulli succeeded in placing it on a sound mathematical foundation, and much that is in the 'Ars Conjectandi' is of value today. Nicolas Bernoulli was familiar with the above treatise and in 1709 he published his 'Specimina Artis Conjectandi ad Quaestiones Juris Applicata,' which is replete with references to his uncle's researches In this work Bernoulli attempted to apply mathematical calculations to various questions relating to the expectation of life.

The next mathematician to cultivate the subject with any degree of success was De Moivre, whose 'Doctrine of Chances, or a Method of Calculating the Probability of Events in Play' appeared in 1718. De Moivre had already communicated a paper to the Royal Society on the subject seven years earlier under the title 'De Mensura Sortis.' His interest in it seems to have been stimulated by a perusal of De Montmort's contribution to the same subject in 1708. In this treatise De Moivre first elaborated the theory of permutations and combinations ; then he proceeded to discuss games of chance depending upon cards and upon dice ; finally he gave solutions of a number of problems found in the treatise of Huygens. The 'Doctrine of Chances' contains seventy-four problems dealing with games of chance as well as a number of questions relating to life insurance. In his demonstrations De Moivre made free use of the method of recurring series (finite differences) ; by the application of this, he was able to solve a number of difficult and complicated problems. In a later work ('Miscellanea Analytica de Seriebus et Quadraturis,' 1730) he investigated problems on duration of play, of which he gave a general solution. By his employment of recurring series and his introduction of the normal distribution curve, De Moivre did more to establish the theory on a firm mathematical foundation than did any other writer with the sole exception of Laplace.

Meanwhile the Bernoulli tradition was being carried on by Daniel. His "Specimen Theoriae Novae de Mensura Sortis," which is to be found in the Commentarii Acad. Petrop. v., 1730–1 (pp. 175–192), is remarkable for its boldness and sagacity. In this work, Bernoulli applied his researches to life insurance and health statistics. He solved some of the problems proposed by Huygens and he gave examples of duration of play, a branch which was later to exercise the highest skill of Lagrange and Laplace. 'Disquisitiones Analyticae de Novo Problemate Conjecturali' (1769) was designed to illustrate the power and the scope of the differential calculus.

'The Nature and Laws of Chance' (1740), from the pen of Thomas Simpson, had some interesting observations on the subject. This is, however, little more than an abridgement of De Moivre's work and there is little in it that was new. Two years later, Simpson published a work on annuities ('Doctrine of Annuities and Reversions,' 1742). Another Englishman, the Rev. Thomas Bayes, turned his attention to the subject. He published in the Phil. Trans. for 1763 (pp. 370–418) a work on probability in which the notion of inverse probability is plainly discernible. Bayes' fundamental theorem may be stated thus: If an event has happened p times and failed q times, the probability that its chance at a single time lies between a and b is

$$\int_a^b x^p(1-x)^q\,dx \div \int_0^1 x^p(1-x)^q\,dx^{(3)}.$$

Other writers who contributed to the subject were the Swiss, Jean Trembley ('Disquisitio Elementaris circa Calculum Probabilium,' 1793), Edward Waring ('Meditationes Algebraicae,' 1770), and Condorcet ('Essai sur l'Application de l'Analyse à la Probabilité,' 1785). The contributions of D'Alembert are not without interest. More than once he had criticized the fundamental principles of the calculus; now he challenged the views of his contemporaries on the subject of probability. His 'Doutes et Questions sur le Calcul des Probabilités,' which is to be found in the 'Mélanges de Philosophie,' I (Paris, 1821), consists mainly of efforts to refute doctrines which were commonly accepted.

The versatile Euler turned his attention to the subject, ("Calcul de la Probabilité dans le Jeu de Rencontre": Hist. de l'Acad. Berlin, 1753, pp. 255–270), but his contributions were of no permanent value. With Lagrange, however, it was different. The development of physical science had now reached such a stage that the application of the doctrine of probability became an important aspect of scientific method. Lagrange's contributions are to be found in the fifth volume of the 'Miscellanea Taurinensia' (1770–83, pp. 167–232). In his exposition he made free use of the calculus; he discussed theories of errors and he first showed how to obtain the best value from a series of non-concordant observations of the same physical quantity ('Mémoire sur l'Utilité de la Méthode de prendre le Milieu entre les Résultats de plusieurs Observations . . . . par le Calcul des Probabilités'). But it was Laplace who made the most outstanding contributions. His masterpiece on this subject did not appear until a dozen years after the close of the period we are considering, but it is clear that the subject had been occupying his attention from about 1770. Some of the memoirs he contributed to the subject are still regarded as classical expositions of the principles. His treatment was purely analytical and he used partial differential equations consistently and extensively. His interest in the subject was stimulated by his researches on celestial mechanics. By his 'Théorie Analytique des Probabilités' (1812) Laplace ranks as the true founder of the theory of probability. Gauss and Legendre also made important contributions; the extensive application of their investigations is not to be detected until the nineteenth century was well established, when Poisson, De Morgan, Sylvester and others turned their attention to it. [See Todhunter: 'History of the Mathematical Theory of Probability,' London, 1865).

4. *Progress in Algebra and Related Subjects.*

In the latter half of the seventeenth century, and during the course of the eighteenth, algebra continued to advance. Although we can detect no outstanding change, it acquired greater flexibility and perfection in all its branches. The construction of cubic and biquadratic equations which Descartes had effected by combining a circle with a parabola (' La Géométrie, Book III,' 1637) was improved. By a skilful application of the binomial theorem Leibniz had succeeded in mastering the so-called irreducible case of Cardan. Euler solved biquadratics by a method which marked an improvement upon that employed by Descartes. His attempts to solve an equation of the fifth degree proved abortive (the impossibility of this was hinted at by Ruffini in 1803, and demonstrated by Abel in 1824). The outstanding developments in the method of solving equations came from Lagrange. He contributed two erudite papers to the Berlin Academy, (' Sur la Résolution des Équations Numériques,' 1767, and ' Reflexions sur la Résolution Algébrique des Équations,' 1770) ; these were followed by his ' Traité de la Résolution des Équations Numériques ' (1798–1808). In these Lagrange made extensive use of the method popularized by Descartes and later employed by Euler and Etienne Bezout ; this consists of finding a resolvent. Lagrange also made important contributions by considering the general conditions of solvability. Fourier also wrote on the solution of numerical equations and, although his results did not appear until 1831 (' Analyse des Équations Determinées '), the substance of this work had been made public as early as 1796 in the course of his lectures at the École Polytechnique. Euler's failure to solve the quintic led to the study of approximate methods of solution ; this branch was successfully prosecuted by Halley and by Taylor.

The nature of equations came in for renewed attention. The formulæ for the computation of the symmetric functions of the roots had been established by Vieta, Harriot, and Newton. A rigorous demonstration of Descartes' Rule of Signs was given by Segner in 1728 and again by de Gua in 1741. Euler added notably to the knowledge of equations by his ' Vollständige Anleitung zur Algebra ' (1770), and so did Waring, an able but obscure analyst, whose ' Meditationes Algebraicae ' appeared in 1770 and in 1782. Étienne Bezout's ' Théorie Générale des Équations Algébriques ' (1779) is a résumé of all that was then known on the subject.

Series were cultivated with vigour. Montmort, in 1714 and again in 1717, and James Bernoulli employed them freely in their investigation of the laws of chance. De Moivre carried these researches much further in the study of the calculus of probability. In 1731 he published an original work explaining the properties of recurring series, already suggested by Daniel Bernoulli in 1728, which he employed freely in his ' Doctrine of Chances.' About the same time, Stirling brought out a treatise on series (' Methodus Differentialis sive Tractatus de Summatione et Interpolatione Serierum Infinitarum,' 1730) in which he considered their convergence, summation and interpolation. Though it is difficult to detect any original plan in his writings, Stirling showed considerable dexterity in transforming one series into another. Euler's work on series led to the development of the integrals known by his name. Though he recognized the need for convergence, he had no satisfactory criterion for detecting it and his somewhat casual employment of infinite series occasionally led him into errors rarely found in a scholar of his sagacity. Waring's views on convergence (1757) were well in advance of his day. He was aware of the well-known test for convergence usually attributed to Cauchy. After Stirling the subject was developed by Simpson, Maclaurin, Landen ('Miscellanea Analytica,' 1762), and later by Pfaff and Krampf. Algebra received a further development by the application of combinatorial analysis, which may be regarded as an important extension of the binomial. Vieta, Mersenne, Schooten, Pascal, Wallis, James Bernoulli, De Moivre, Euler, had already contributed to this subject. The greatest expansion, however, came from Germany. Hindenburg published a tract in 1779 which threw much light upon the subject.

Many writers of the period busied themselves with the compilation of treatises on algebra and some of these reached a very high standard. Euler's 'Algebra' has already been mentioned. Though not free from blemish it was a work of considerable merit. Hardly less important were Maclaurin's 'Elements of Algebra' (1748), which, however, was left incomplete, and Clairaut's 'Élémens d'Algèbre' (1749 and 1760). Vandermonde contributed a work in which occurs the first systematic exposition of determinants. He also gave the theorem still known by his name. Simpson's 'Elementary Treatise on Algebra' (1745), Sanderson's 'Treatise on Algebra' (1740), Fantet de Lagny's 'Nouveaux Élémens d'Algèbre' (1727) are important as reflecting the interest which was now being devoted to the subject.

The subject of imaginaries began to attract increasing attention. They first appear in Cardan ('Ars Magna,' 1545) and again in Bombelli, Girard and Wallis. Leibniz developed the subject in his study of Cardan's rules when he showed $\sqrt{(1+\sqrt{-3})}+\sqrt{(1-\sqrt{-3})}$ to be equal to $\sqrt{6}$. Newton's work on imaginaries is confined to his consideration of the number of roots of an equation. Fagnano gave an expression for the value of π which involved the use of imaginaries. Some attempt at graphic representation is faintly discernible in Wallis ('Algebra,' 1685). It was, however, Caspar Wessel who, in 'An Essay on the Analytical Representation of Direction' (1797), first gave the modern treatment. Argand extended the subject in 1806, but it was left for Gauss to establish the principles clearly and systematically. In his dissertation for the doctorate (1799) Gauss showed that every algebraic equation has a root of the form $(a+b\sqrt{-1})$. The introduction of i for $\sqrt{-1}$ is due to Euler. Cauchyu sed imaginaries freely; to him are due some of the improvements in notation which are still current.

D'Alembert (Mem. de l'Acad. Berlin, 1747) showed that imaginary quantities could be reduced to the form $a+b\sqrt{-1}$, a and b being real. De Moivre and Cotes had already detected the way in which $\sqrt{-1}$ provides a link between the logarithmic and trigonometric functions.

The theory of numbers, inseparably associated with the name of Fermat, was revived by Euler. The second volume of his 'Algebra' contains solutions of some of the problems proposed by Fermat, *e. g.*, every number whatsoever consists of not more than four squares. Waring ('Meditationes Algebraicae,' 1770) also has some contributions on the subject. But the most notable advances came from Lagrange. These are contained in the third volume of the 'Miscellanea Taurinensia.' Several of his earlier memoirs contain solutions of problems upon this subject, for which he appears to have had a peculiar gift. The subject was taken up by Gauss ('Disquisitiones Arithmeticae,' 1801) and later by Legendre, whose 'Théorie des Nombres' (1830) remains a standard work on the subject. This contains the first satisfactory proof of the theorem of quadratic reciprocity. Legendre treated the subject as a branch of algebra and he carried it as far as the current algebraical methods allowed. The researches of Gauss and Legendre provided a starting point for the investigations of Jacobi, Dirichlet, Riemann and others during the nineteenth century.

5. *Progress in Geometry and Trigonometry.*

As we have already noted, progress in the new geometrical methods was slow during the latter half of the seventeenth century. The first exposition of the science of coordinate geometry was Descartes' 'La Géométrie' (1637). In this work Descartes deliberately left many obscurities which made it unintelligible to many. By the time De Beaune and van Schooten had elucidated these the new calculus was well established, and for nearly a century mathematicians were too busy exploring the new treasures unfolded before their eyes to pay much heed to the seemingly less fruitful geometrical methods. Nevertheless the possibilities of the subject were recognized by a few writers later in the century; of these the most important were Jan de Witt ('Elementa Curvarum Linearum,' 1659) and La Hire ('Construction des Équations Analytiques,' 1679). Newton developed the subject still further in his "Enumeratio Linearum tertii Ordinis," 1668 ('Opuscula Newtoni,' I, pp. 235–270) and, by the time the 'Principia' appeared, the main principles were well established. Newton's contributions were still further advanced by Stirling, whose 'Lineae tertii Ordinis Neutonianae sive Illustratio Tractatus D. Neutoni de Enumeratione Linearum tertii Ordinis' appeared in 1717. With the publication of this work the new methods may be said to have come into their own.

Descartes, in his treatise, had restricted his exposition to curves of two dimensions. At the end of the second book of 'La Géométrie' is a hint—the only one in all his writings—of the possibility of extending the investigation to three dimensions. He did not pursue the matter, and it was not until the turn of the century that the hint was followed up. In 1700 (July 24th and Aug. 23rd) Antoine Parent submitted a paper to the Académie des Sciences (" Des Affections des Superficies ") in which Descartes' principles were extended to solids. This paper, which is included in Parent's ' Essais et Recherches de Mathématiques et Physiques ' (1713), proved an inspiration to Clairaut, whose ' Recherches sur les Courbes à Double Courbure ' appeared in 1731. Substantial contributions were also made by De Gua " Démonstration de la Règle de Descartes " (Hist. de l'Acad. des Sciences, 1741, pp. 72–96), but the next important development came from Euler in his ' Introductio in Analysin Infinitorum,' 1748. The second part of this work deals with analytical geometry both plane and solid, each being treated with complete mastery. Euler investigated tangents and tangent planes, normals, areas and volumes, and he gave an exhaustive account of curves defined by equations of the second degree. In the same year there appeared a work hardly less important. This was the ' Instituzioni Analitiche ad uso della Gioventù Italiana ' (1748) of Maria Agnesi. This was translated into French in 1755 and into English (' Analytical Institution,' Colson) in 1801. The ' Introduction à l'Analyse des Lignes Courbes algébriques ' (1750) of Gabriel Cramer was also a work of merit, and so was an earlier work, ' Nouvelles Pensées sur le Système de M. Descartes ' (1730), from the pen of John Bernoulli.

Meanwhile the projective geometry of Desargues had suffered an even more humiliating fate. Published in 1639, the ' Brouillon Proiect ' contains the fundamental theorems on involution, poles and polar, and the theory of perspective which Desargues employed to investigate the properties of conics. Desargues' methods seemed to have impressed Pascal, whose ' Essay pour les Coniques ' (1639) appeared when he was but sixteen. Despite this, however, nearly a century was to elapse before projective geometry again began to attract notice. The revival began with the publication of Maclaurin's ' Geometria Organica ' (1720), which contained an elegant investigation of the curves of the second order by regarding them as generated by the conditional intersection of lines and angles turning about fixed points or *poles*. But the real extension did not come until a group of French mathematicians, Monge, Carnot, Poncelet, turned their attention to the subject. Monge is usually regarded as the inventor of descriptive geometry. In his ' Géométrie Descriptive, Leçons données a l'École Normale 1794–95 ' (Paris, 1799) he elaborated the main principles. He developed the theory of perspective and he derived the properties of surfaces from perpendicular projections on to three surfaces. He also solved many important problems relating to the intersection of planes. Monge also contributed a number of papers on the same subject ; many of these are embodied in his ' Application de l'Algèbre à la Géométrie ' (2nd edn., 1850). Carnot, in his ' Géométrie de Position ' (1803), opened up new prospects. He also revived and extended the work of Maclaurin under the title of transversals (' Essai sur les Transversals,' 1806). Poncelet turned aside from analysis and infused new vigour into purely geometrical methods in his ' Traité des Propriétés Projectives des Figures ' (1822). Many of his results had already appeared in Crelle's ' Journal.' Poncelet freely employed projection and reciprocation to deduce many important properties of figures.

In elementary geometry there had been little development since the Greeks. In the eighteenth century many writers published works founded on the ancient geometry. David Gregory brought out an edition of Euclid in 1703. Thomas Simpson's ' Elements of Plane Geometry ' appeared in 1747; this was followed by Robert Simson's ' Elements of Euclid ' in 1756, and later by Clairaut's ' Élémens de Géométrie ' in 1765. Euler gave a proof of the theorem giving the relation between the faces, vertices and edges of a polyhedron ($E+2=F+V$); this, however, occurs in a little-known work of Descartes. Of considerable importance was the fact that the foundations of the subject began to be critically studied and serious attempts were made to prove the famous parallel postulate of Euclid. Gauss maintained, and Beltrami subsequently demonstrated, that the postulate could not be proved. This led to the development of non-Euclidean geometries. The first contributions appear to have been made by Saccheri (' Euclides ab omni naevo Vindicatus,' 1733), a work from which Lobachevsky no doubt drew inspiration for his ' Ueber die Principien der Geometrie ' in 1829. Legendre's ' Élémens de Géométrie ' (1794) was designed to replace the ' Elements ' of Euclid.

We now turn to the developments in trigonometry. During the eighteenth century trigonometry was simplified and considerably extended. This was due in no small measure to the improvements in notation which had been made in the seventeenth. Oughtred (' Trigonometrie,' 1657), Caswell (whose contributions are to be found in Wallis's ' Algebra,' 1685), Newton and Cotes all shared in this development. Newton gave formulæ for $\sin nx$ and $\cos nx$ and the Frenchman de Lagny did the same thing for $\tan nx$ and $\sec nx$. John Bernoulli treated trigonometry by analytical methods and his son Daniel did useful work in computing the trigonometrical functions. He was the first to employ the present notation for the inverse trigonometrical functions (1729). Euler (1798) gave the formula for $\tan^{-1} x$ which was an improvement upon that by James Gregory. But the main developments lay in the direction of the use of imaginaries in trigonometry. Cotes (Phil. Trans., 1714, and ' Harmonia Mensurarum,' 1722) showed that $ix=\log(\cos x+i\sin x)$. De Moivre established the theorem known by his name, viz., that $\cos nx+i\sin nx$ is one value of $(\cos x+i\sin x)^n$, n being rational. Euler in his ' Introductio in Analysin Infinitorum ' (1748), established the relation $e^{ix}=\cos x+i\sin x$. The work also contains a number of formulæ in spherical triangles. Euler also introduced the notion of indicating the sides of a triangle by the letters a, b, c.

Lagrange, in 1761 gave a rigorous proof that π is irrational, so also did Legendre. In 1768 the Prussian, Johann Heinrich Lambert, published a paper in which he developed the results established by De Moivre. In this he introduced the hyperbolic sine and cosine, though he did not employ the symbols $\sinh x$, $\cosh x$ until three years later. Gauss, in a work on celestial mechanics ('Theoria Motus Corporum Coelestium,' 1809), obtained the formulæ now known as Gauss's Analogies, which are used in spherical trigonometry. Simpson ('Trigonometry, Plane and Spherical,' 1748), and later La Croix, wrote on trigonometry, but there is nothing original in their writings.

B. Developments in Applied Mathematics.

The progress made in applied mathematics during the seventeenth century was in some respects even more spectacular than that in pure mathematics. Yet the two advances were intimately related. The increased attention to mechanics during the latter half of the century reacted powerfully upon the development of mathematics. It had been a period of great experimental activity and this had given rise to the appearance of fresh problems in statics and, more particularly, dynamics. But the dawn of the eighteenth century found men lacking the mathematical equipment needed to cope with the problems raised. To supply this need was the task to which men like Euler, D'Alembert, Lagrange, applied themselves. Not the least important aspect of their work lies in the elaboration of principles and methods by which many of the problems bequeathed by their predecessors could be successfully attacked.

During the eighteen hundred years which separated Archimedes and Stevin, progress in mechanics had been unaccountably slow. The elucidation of the fundamental problems of statics and dynamics attracted the attention of singularly few men of science, and even these seemed to be content with nothing more than prolix commentaries upon the writings of Aristotle. Some few indeed did turn to the subject; such are Nicolas of Cusa, Cardan, Tartaglia and others, but their writings reveal little originality. The researches of Stevin had opened up a new world and in his 'Hypomnemata Mathematica' (1608) he had established the important principle of the composition of forces. In the meantime, whilst Francis Bacon was philosophizing about the new experimental methods, Galileo was putting them into practice, and by his study of the motion of falling bodies had already challenged the omniscience of the great Stagyrite philosopher. The researches of Galileo had stimulated others—Castelli, Viviani, Baliani in Italy, and Roberval and Mersenne in France—to turn their attention to this important subject. Descartes, in an attempt to make his vortex theory a workable system ('Principia Philosophiae,' Part II, 1644), had devised a completely new system of mechanics. But it was founded upon an entirely erroneous conception of motion and it did nothing to advance the subject. Much more enduring were the contributions of John Wallis, whose 'Mechanica sive De Motu,

Tractatus Geometricus' (1670) did much to establish the subject upon firm foundations and thus blaze the trail for Newton. With the publication of the first two books of the 'Principia' it is not too much to say that Newton transformed mechanics into an entirely new science.

The investigation of the principles relating to the composition of forces and of the laws governing motion provided a convenient starting point for the development of mechanics during the eighteenth century. Pierre Varignon had studied the composition of forces about 1687 and had simplified the proofs of many of the theorems already established. John Bernoulli furthered the science of statics by proposing the principle of virtual work which Varignon had already applied in his 'Nouvelle Mécanique ou Statique' (1725). Bernoulli also gave greater generality to the important principle of the conservation of living forces (*vires vivae*) suggested earlier by Huygens. By noting that the same body can receive different gravitational accelerations, Bernoulli was first led to the distinction between mass and weight ('Meditatio de Natura Centri Oscillationis: Opera Omnia,' II, p. 168)[(4)]. James Bernoulli contributed to statics by first giving a demonstration of the form assumed by a hanging chain, a problem which had already been considered by Galileo. He also studied the behaviour of beams under various loads and he tried to establish a relation between deflection and load. Another of this gifted family, Daniel, gave one of the earliest demonstrations of the parallelogram of forces. He also discovered, simultaneously with Euler, the principle of the conservation of areas (1746).

An important contribution was made by Maclaurin ('Complete Treatise of Fluxions,' 1742). He employed three mutually perpendicular axes for the resolution of forces. This was seized upon by D'Alembert, who devised a means of reducing dynamical problems to problems in statics. Of the other great mathematicians of the period, Euler, Clairaut, Lagrange, Laplace, etc., their contributions were mainly to dynamics; nevertheless, they enriched the science of statics by considering special problems, as for example, elasticity, the vibration of strings, etc. In the investigation of the latter problem the names of Taylor, John and Daniel Bernoulli, D'Alembert and Euler are conspicuous. Euler also devoted some attention to the bending of beams, and although he left many things unexplained, his work forms an important link between that of Daniel Bernoulli and Coulomb, to whom much of the modern theory is due. Monge, better known for his geometrical researches, published a treatise on 'Statics' in 1786 which exhibits his usual thoroughness and skill.

Dynamics proved a much more fruitful source of enquiry. The law of falling bodies had been correctly enunciated by Baliani in his 'De Motu Naturali Gravium Solidorum et Liquidorum,' Prop. V (Genoa, 2nd Edn., 1646). Eight years earlier, Galileo's 'Discorsi e Dimostrazioni Matematiche' had initiated a complete reform in what had hitherto been a neglected branch of science. Nearly fifty years later Huygens had extended Galileo's investigation to motion in a circle. His pendulum experiments led James Bernoulli to attempt to explain the compound pendulum on the principle of the lever, but his observations were not free from error. In his papers to the 'Acta Eruditorum' (1691) and the Proc. Paris Acad. (1703), he appears to have been more successful. His brother John also attacked the question. More important, however, were the latter's investigation of the impact of bodies. Following Descartes' observations on the subject, in which he rarely stumbled across the truth, Wallis, Wren and Huygens had made contributions of value. In 1724 the Académie des Sciences proposed for their prize essay the laws of impact of bodies. John Bernoulli's solution is contained in a work 'Discours sur les Loix de la Communication du Mouvement,' which was written in the style of Leibniz. But it was a disciple and fellow-countryman of the Bernoullis, J. Hermann, who produced the first treatise on dynamics fashioned upon the analytical model. This was his 'Phoronomia' (1716). Despite a certain clarity and conciseness, the work lacks completeness and it is not distinguished by any great novelty of conception.

The task of compiling a work on dynamics, complete and original in all its branches, fell to Euler who directed his energies to the production of his 'Mechanica, sive Motus Scientia Analytice Exposita' (1736). Earlier contributions to mechanics from his pen are to be found in the Transactions of the St. Petersburg Academy. Euler considered that the new problems that were arising were beyond the scope of the geometrical methods favoured by Newton and his followers, and he tried to replace them by the more powerful analytical methods. He was not entirely successful. That was left for Lagrange. But by his exposition, Euler certainly advanced a long way in that direction. He also enriched dynamics with the equations still known by his name concerning the motion of a rigid body about a fixed axis. These are to be found in the Memoirs of the Royal Academy of Science, Berlin, 1718. It has been said, and not without reason, that the 'Mechanica' of Euler did for mechanics what 'La Géométrie' of Descartes did for geometry.

In 1743 there appeared D'Alembert's 'Traité de Dynamique,' an important landmark in the development of the subject. Like almost every other prominent mathematician of the period D'Alembert studied the pendulum, and his researches on the centre of oscillation led him to a series of simple observations which ultimately he generalized and embodied in the principle which goes by his name, viz., that the internal actions and reactions of any system of rigid bodies in motion balance. Following Bernoulli, D'Alembert showed that in any system of bodies acting mutually their several motions at any instant are decomposed into two portions, one of which is retained in the next instant and the other spent, and since an equilibrium must obtain among the lost motions an expression is hence derived for the motions that are preserved. By the application of this principle the most complex problems in dynamics are thus reducible to simple statical problems and solved by generalized methods. D'Alembert, by use of this principle, gave an elegant and complete solution in 1747 of the difficult problem of the precession of the equinoxes. He also attacked the *three bodies* problem. The application of his method was further extended by Maclaurin's use of coordinates to represent forces. Later (1744 and 1751), D'Alembert extended his principle to hydrodynamics.

About this time George Atwood studied the laws of falling bodies. In 1784 he published a description of Atwood's Machine in a treatise 'On the Rectilinear Motion and Rotation of Bodies.'

Clairaut, whose main contributions lie in his improvement of the mathematical technique, also devoted some attention to celestial mechanics. He investigated the form which a mass of fluid rotating about its centre of mass would assume. His 'Théorie de la Figure de la Terre' (1743) contains his formula for the acceleration due to gravity at different latitudes. In a later work, 'Théorie de la Lune' (1752), Clairaut abandoned analysis in favour of the geometrical methods of Newton and Maclaurin. Clairaut's work was taken up by Boscovitch, whose 'Theory of the Constitution of the Universe,' purged of its antiquated metaphysics, contains some contributions of importance. He postulated that atoms were nothing more than centres of attractive and repulsive forces acting at a distance. Another writer of this period whose works deserve notice is Segner, an original thinker who occupied the Chair of Mathematics at Göttingen. In 1755 he published a short essay on rotating bodies ('Specimina Theoriae Turbinum'), in which are contributions of importance to physical astronomy. It was Segner's invention of the water wheel which led Euler to the study of machines operated by running water.

The science of dynamics reached its highest point of perfection in the 'Mécanique Analytique' of Lagrange (1788). In this work Lagrange combined D'Alembert's principle with that of virtual velocities and thus converted the whole into an absolute analytical system of great power and flexibility. His method was to refer the efforts of every particle of a moving system to three mutually perpendicular axes and thence derive three separate differential equations which on integration would give the final solution of the problem. By applying the calculus of variations he showed how the whole of statics could be reduced to a single principle, that of virtual work. Although his main contributions are contained in the monumental work referred to above, there are numerous memoirs in the 'Miscellanea Taurinensia' (Vol. III), in which Lagrange illustrates the generality of his methods by a number of examples. Lagrange also made an attack on the celebrated *three bodies* problem; in this, the motion of three bodies each of which attracts the other two in accordance with the law of gravitation is investigated. For his 'Essai sur le Problème des trois Corps' Lagrange was awarded the prize offered by the Paris Academy in 1772. The same problem was attacked by Laplace, who succeeded in obtaining the general equations of motion. Laplace was interested primarily in celestial mechanics, which overshadowed all his other work. In his great treatise, 'Mécanique Céleste' (1799–1825), he determined the attraction of a spheroid on a particle outside it, an investigation which had already been foreshadowed by Legendre. Laplace introduced the idea of spherical harmonic analysis. The notion of potential, which was to play such a vital part in nineteenth century physical science, was also developed by him. His researches made possible an attack on physical problems which otherwise would have been unapproachable.

Projectiles.—Besides the general developments considered above, certain specific problems attracted the notice of many writers. One of the most important of these was the motion of projectiles. Galileo and others, neglecting the resistance of the air, had shown the path of a projectile to be a parabola. Galileo was never of the opinion that the resistance of the air played no part in the motion; he had no means of attacking the problem mathematically. The contributions of John Wallis are important if only because of the emphasis he placed upon an appeal to experiment. With his usual thoroughness he tried to place the fundamental notions on a sound basis. He was agreed that "the random of a bullet is very different from that of a parabola," and he promised, in response to a request from Halley, to investigate the problem and communicate his results to the Royal Society. In this he was forestalled by Newton.

Newton declared that the resistance offered to a projectile moving through the air was proportional to the square of its speed. After Newton, John Bernoulli was the next to pay serious attention to the problem. Although he, too, lacked the necessary mathematical ability, he nevertheless, by a wise combination of calculation and experiment, obtained some conclusions of value. The subject was taken up by Benjamin Robins who, by his 'New Principles of Gunnery' (1742), added notably to the theory of the flight of projectiles. In the course of his researches he devised an ingenious method of determining the impulse of a shot by firing it against a heavy loaded pendulum, the deflection of which correctly indicated the momentum thus communicated. From his observations he was able to deduce the velocity of the shot at any stage of its flight. He concluded that Newton's statement that the resistance was proportional to the square of the velocity was only true for low velocities. At higher velocities the resistance increased at a much faster rate. Robins' ballistic pendulum was adapted by Hutton, Professor of Mathematics at the Royal Military Academy, Woolwich, in an elaborate series of experiments on the velocity of rifle bullets, in 1790 and 1791.

The investigation of the motion of projectiles was taken up by an able geometer, J. C. Borda, who tried to determine the nature of the curve described by a projectile moving through a resisting medium, and also to determine the elevation needed to produce the maximum range. His results are contained in the Hist. de l'Acad., 1769 (p. 121). He found that the elevation to produce the maximum range depended upon several factors, *e. g.*, weight of the projectile, its initial velocity, but that in no case was it 45°. A projectile of weight 140 pounds, with an initial velocity of 600 feet per second, was found to have a maximum range when the elevation was 37° 15′ ; a similar projectile with an initial velocity of 1100 feet per second produced a maximum range at 33° 20′. Borda's investigations on the motion of projectiles were continued by Templehof in 1781.

Euler, in an Addendum to his 'Methodus Inveniendi Lineas Curvas,' studied the question of the motion of projectiles (' De Motu Projectorum '). In this he considered the motion of a particle moving in a resisting medium. In his treatment Euler employed the calculus of variations, and his investigation led to an important principle which was further developed later in the century. This was the Principle of Least Action. The origin of this principle is to be found in antiquity when Euclid and others assumed the rectilinear propagation of light. Hero of Alexandria showed that in reflection at plane surfaces the path taken by a ray of light was a minimum. Nearly fifteen centuries later Fermat showed that a ray of light passing from a point in one medium to a point in another would so adjust its path that it described it in the least possible time. It was, however, Maupertuis who brought the subject of Least Action into prominence. He declared (' Recherches des Loix du Mouvement ' : Acad. Roy. des Sciences, Apl. 15, 1744) that whenever any change occurs in nature the quantity of action employed in this change is always the least possible, and he maintained that from this one fundamental principle all the laws of mechanics could be deduced. The principle as enunciated was not free from obscurities, and led to endless disputes as to what was involved in the term *action*, Maupertuis maintaining that it was the continued product of mass, velocity and space. Eventually these were clarified during the nineteenth century, mainly by Hamilton, and the theory has since played an increasingly important part in the development of physical science.

Side by side with the rapid advances which had been made in statics and dynamics is to be observed an awakened interest in hydrodynamics, or the science of the motion of fluids. The conditions of equilibrium of a fluid had been discovered by Archimedes but little further was contributed until Stevin traced the effects of fluid pressure and obtained results which were subsequently verified by Pascal (' Récit de la Grande Expérience de l'Équilibre des Liqueurs,' Paris, 1648). Pascal's treatise is probably the first in which the laws of hydrostatics are simply and clearly demonstrated. The genius of Galileo is reflected in two of his pupils, Castelli, whose ' Della Misura dell'Acque Correnti ' (Rome, 1628) created a new department of hydraulics, and Torricelli, whose ' Opera Geometrica ' (Firenza, 1644) is remarkable for some noteworthy contributions to the science of hydrostatics.

There were two problems in hydrostatics which occupied the attention of the mathematicians of the eighteenth century. These were, first, the resistance encountered by a solid moving through a fluid, and secondly, the efflux of liquids from an aperture in the base of the containing vessel.

The first of these appears to have been first undertaken by Torricelli in the treatise mentioned above. His work was mainly experimental. Newton undertook the examination of Torricelli's results in the second part of the 'Principia,' but he does not appear to have been entirely satisfied with his conclusions.

Daniel Bernoulli ('Exercitationes quaedam Mathematicae,' Padua, 1724), investigated the motion of fluids by the principle of *vis viva.* He studied the problem by a series of experiments in which he observed the flow of water in canals, the width and depth of which could be varied. Meanwhile, his father ('Joh: Bernoulli. Hydraulica nunc primum detecta et directe demonstrata ex principiis pure Mecanicis, anno 1732') made observations on the same problem.

Clairaut, in his treatise on the figure of the earth (1743), contributed to the subject; so did Euler and D'Alembert. The former, in a series of memoirs 1755 to 1772, reduced the subject to strictly analytical form. D'Alembert's contribution was his 'Traité de l'Équilibre et du Mouvement des Fluides' (1744). In 1750 the problem of the theory of the resistance of fluids was proposed by the Berlin Academy. To the questions raised, D'Alembert ('Essai d'une Nouvelle Théorie sur la Résistance des Fluides,' 1752) submitted a general solution in which he tried to found the subject of sound mathematical principles with the aid of analytical methods. His work was taken up some years later by Condorcet and by Bossut. The latter published the results of some observations on the flow of water in tubes ('Traité d'Hydrodynamique,' 1786). Condorcet made observations on the resistance encountered by vessels dragged through water by forces of known magnitude.

Meanwhile, Lagrange in 1781 examined the chief difficulties which surrounded the subject. Seven years later, in his 'Mécanique Analytique,' he still further simplified and generalized the subject. He united the two branches, the mechanics of solids and the mechanics of fluids, which had hitherto pursued separate courses, into one whole. In the first part of his treatise the problems relating to the equilibrium of bodies, fluids as well as solids are treated as branches of one and the same science and brought within the scope of his powerful analytical methods.

The investigation of the discharge of fluids had exercised the sagacity of Newton, who studied the flow of water through a small hole in the base of a cylinder. Twenty years later Varignon attacked the same problem. But a more complete investigation came from Daniel Bernoulli ('Hydrodynamica, seu de viribus et motibus Fluidorum Commentarii,' 1738). His work was largely experimental, and the principle of *vis viva* is freely employed. D'Alembert, in his treatise on the equilibrium and motion of fluids (1744), also studied the question.

But it was in Italy that the problem attracted the greatest notice. Michelotti, a distinguished Italian doctor, described his observations in a memoir: 'Sur la détermination des vitesses de l'eau, etc. (Mem. Torino vii.). In this he rejected the conclusions already reached by Newton. Others who contributed to the subject were Gabrielle Manfredi, who enriched with excellent notes the edition of Guglielmini's 'Della Natura dei Fiumi,' and Giovanni Poleni, Professor of Astronomy at Padua. His 'De Castellis per quae derivantur Fluviorum Aquae,' which was published in Padua in 1717, is especially important. To him also is due a new edition of Frontinus, entitled: 'Frontini, de aquae ductibus Urbis Romae Commentarius,' to which he added valuable notes.

References.

(1) Dict. Nat. Biog. lix. p. 384.
(2) Wolf, 'History of Science, Technology and Philosophy in 18th Century' (p. 52).
(3) Cajori, 'History of Mathematics' (1924), p. 230.
(4) Mach, 'Science of Mechanics' (p. 251).

List of Works Consulted.

Ball, W. W. R., 'Short Account of the History of Mathematics,' 1927.
Bell, E. T., 'Development of Mathematics,' 1945.
Cajori, F., 'A History of Mathematics,' 1924.
Cantor, M., *Vorlesungen über Geschichte der Mathematik*, Vol. III, 1898.
Dissertation Fourth. Encyclopædia Brit., 7th Edn.
Wolf, A., 'History of Science, Technology, and Philosophy in the 18th Century,' 1938.

List of Writers Mentioned, with Dates.

Abel, N. H. (1802–1829).
Agnesi, M. G. (1718–1799).
Argand, J. R. (1768–1822).
Atwood, G. (1746–1807).
Bacon, F. (1561–1626).
Baliani, G. B. (1582 or 6–1660 or 6).
Bayes, T. (? –1761).
Beaune, F. de (1601–1652).
Beltrami, E. (1835–1900).
Berkeley, G. (1685–1753).
Bernoulli, D. (1700–1782).
Bernoulli, Jas. (1654–1705).
Bernoulli, John (1667–1748).
Bernoulli, Nicolas (1687–1759).
Bézout, E. (1730–1783).
Bombelli, R. (16th cent.).
Borda, J. C. (end of 18th cent.).
Boscovitch, R. (1711–1787).
Bossut, C. (1730–1814).
Braikenridge, W. (1700–1759).
Cardan, G. (1501–1576).
Carnot, L. N. M. (1753–1823).
Castelli, B. (1577–1644).
Caswell, J. (fl. 1680–1700).
Cauchy, A. L. (1789–1857).
Ceva, G. (1647–1734).
Clairaut, A. C. (1713–1765).
Condorcet, N. C. de (1743–1794).
Cotes, R. (1682–1716).
Coulomb, C. A. de (1736–1806).
Cramer, G. (1704–1752).
Cusa, N. (1401–1464).
D'Alembert, J. (1717–1783).
De Moivre, A. (1667–1754).
De Morgan, A. (1806–1871).
Desargues, G. (1593–1662).
Descartes, R. (1596–1650).
Dirichlet, P. (1805–1859).
Euler, L. (1707–1783).
Fermat, P. (1601–1665).
Fagnano, G. (1682–1766).
Fantet de Langy, T. (1660–1734).
Galileo, G. (1564–1642).
Gauss, K. F. (1777–1855).
Girard, A. (1590–1633).
Gregory, D. (1661–1708).
Gua, J. P. de (1713–1785).
Halley, E. (1656–1742).

Hamilton, W. R. (1805–1865).
Hermann, J. (1678–1733).
Hindenburg, C. F. (1741–1808).
Hutton, C. (1737–1823).
Jacobi, K. F. A. (1804–1851).
Krampf, C. (1760–1826).
La Croix, S. F. (1765–1843).
Lagny, Thomas Fantet de (1660–1734).
Lagrange, J. L. (1736–1813).
La Hire, P. (1640–1718).
Lalande, J. J. (1732–1807).
Lambert, J. H. (1728–1777).
Landen, J. (1719–1790).
Laplace, P. S. (1749–1827).
Legendre, A. M. (1752–1833).
Leibniz, G. (1646–1716).
L'Hôpital, G. F. A. (1661–1704).
Lobachevsky, N. L. (1728–1777).
Maclaurin, C. (1698–1746).
Manfredi, G. (1681–1761).
Maupertuis, P. L. M. de (1698–1759).
Mersenne, M. (1588–1648).
Michelotti, I. (1764–1846).
Monge, G. (1746–1818).
Montmort, P. R. (1678–1719).
Newton, I. (1642–1727).
Oughtred, W. (1574–1660).
Pacioli, L. (1445–1514).
Parent, A. (1666–1716).
Pascal, B. (1623–1662).
Pfaff, J. F. (1765–1825).
Poisson, S. D. (1781–1840).
Poleni, G. (1683–1761).
Poncelet, J. V. (1788–1867).
Riccati, Count (1676–1754).
Riccati, V. (1707–1775).
Riemann, G. F. B. (1826–1866).
Roberval, G. P. de (1602–1675).
Robins, B. (1707–1751).
Ruffini, P. (1765–1822).
Saccheri, G. (1667–1733).
Schooten, F. van (1584–1641).
Segner, J. A. (1704–1777).
Simpson, T. (1710–1761).
Simson, R. (1687–1768).
Stevin, S. (1548–1620).
Stirling, J. (1696–1770).
Sylvester, J. J. (1814–1897).
Tartaglia, N. (1500–1557).
Taylor, B. (1685–1731).
Trembley, J. (1791–1811).
Torricelli, E. (1608–1647).
Varignon, P. (1654–1722).
Vandermonde, A. T. (1735–1796).
Vieta, F. (1540–1603).
Viviani, V. (1622–1703).
Wallis, J. (1616–1703).
Waring, E. (1734–1798).
Wessell, C. (1745–1818).
Weierstrass, K. (1815–1897).
Witt, Jan de (1625–1672).
Wren, C. (1632–1723).

ENGINEERING AND INVENTION IN THE EIGHTEENTH CENTURY

By ENGINEER CAPTAIN EDGAR C. SMITH, O.B.E., R.N.

I.

WHEN the first issues of the Philosophical Magazine appeared, the effects of Tudor policy in fostering trade and manufactures had shown themselves plainly; Great Britain had become the foremost commercial nation in the world and the country was fast becoming a great workshop. Many enterprising and ingenious men, some of foreign origin—merchants, manufacturers, inventors and engineers—had contributed to the remarkable progress of the century and their work was to be seen in a hundred directions. The population, which at the time of Elizabeth had been about five million, had at least trebled, and our mines, mills, factories and workshops were active as never before. Communications and transport had been improved, commerce both at home and overseas had reached new levels, and both civil and mechanical engineering had become recognized professions. Such was the wealth created by the combined activities of the people that Lecky wrote that "the triumphant issue of the Great French War was largely, if not mainly due to the cotton mill and the steam engine. England might well place the statues of Watt and Arkwright by the side of Wellington and Nelson for had it not been for the wealth they created she could never have supported an expenditure which during the last ten years of the war averaged more than £84,000,000, nor could she have endured without bankruptcy a national debt which had risen in 1816 to £884,000,000." The national wealth might have been still greater had not so much been squandered on a foolish war, for as Benjamin Franklin wrote to Sir Joseph Banks in 1783, "I join with you most cordially in rejoicing at the return of peace . . . What vast additions to the convenience and comforts of living might mankind have acquired if the money spent in wars had been employed in works of public utility. What an extension of agriculture, even to the tops of our mountains, what rivers rendered navigable or joined by canals, what bridges, aqueducts, new roads, and other public works, edifices and improvements, rendering England a complete paradise, might have been obtained by spending those millions in doing good, which in the last war have been spent in doing mischief, in bringing misery into thousands of families and destroying the lives of many thousands of working people who might have performed the useful labour."

The history of any century can be viewed from many angles and the eighteenth century is no exception. The century of Newcomen, the Darbys, Brindley, Smeaton, Arkwright, Wedgwood, Boulton and Watt was certainly a century of material progress, but it is notable for numerous other reasons. It was a century of famous land and sea battles, and in between the victories of Marlborough and Nelson were the events of the Seven Years War, the campaigns of Clive and Hastings in India, and the struggle in North America. One notable feature of the time was the spread of colonization and the discoveries of explorers, among whom none stands higher than Captain James Cook. It was also a century of political and social progress. If corruption and rotten boroughs existed there was an antidote in the liberty of the press and persons. Side by side with widespread indifference, scepticism and cynicism there was a general toleration in politics and religion, likewise a revival of the latter and a genuine philanthropy. John Wesley's long and fruitful life covered three-quarters of the century, the end of which saw the self-sacrificing labours of John Howard for the reform of prisons and the work of Clarkson and Wilberforce for the abolition of the slave trade, although England had gained much wealth thereby. The callousness of the times was reflected in the terrible punishments, the crowds at executions, and the public exhibition of the unhappy inmates of Bethlehem Hospital. As late as 1774 the 'Morning Post' reported the burning at the stake of a woman. If a glance at society does not reveal the same great contrasts between the luxury and extravagance of the few and the hardships and misery of the many, as was to be seen in France, notorious evils and abuses were common. As for education, this was for the fortunate only, but even they were condemned to the study of dead languages. For the generality of children there was little but drudgery from early childhood. Of schoolmasters, Steele, in one of his essays wrote: "I have very often with much sorrow bewailed the misfortune of the children of Great Britain, when I consider the ignorance and undiscerning of the generality of schoolmasters . . . Many of these stupid tyrants exercise their cruelty without any manner of distinction of the capacities of children, or the intentions of parents in their behalf." Few, indeed, of the pioneers in engineering and industry during the eighteenth century owed much to their educators.

Yet withal it was a period of great literature. A century, the beginning of which could count an Addison, a Pope and a Swift, and at the end could boast of Wordsworth, Coleridge and Lamb, and which saw the writings of Fielding, Hume, Johnson, Gibbon, Bentham and Adam Smith, must be counted a great century. Then, too, newspapers and periodicals multiplied and literary and scientific societies came into being in Edinburgh, Dublin, Newcastle, Leeds and Manchester. The present Royal Society of Arts and Manufactures was founded in 1758, and Rumford's Royal Institution in the closing years of the century. Among the many periodicals published, of special interest here, was the Ladies' Diary, a forerunner of the Philosophical Magazine. It was both an almanac and a scientific journal and was contributed to by many men afterwards distinguished. Founded by John Tipper in 1704, on his death in 1713 it was edited in turn by Henry Beighton, Robert Heath, Thomas Simpson, Edward Rollinson and Charles Hutton. Beighton was one of the improvers of the steam engine while Simpson and Hutton both held successively the chair of mathematics at the Royal Military Academy, Woolwich. The Gentleman's Diary, first published in 1741, was amalgamated with the Ladies' Diary a hundred years later, and the combined magazine continued to appear till 1871.

Many travellers of the eighteenth century left accounts of their journeys through England and from their writings can be gained pictures of the conditions of our roads and water ways, the growth of our manufactures and the state of agriculture. One of these was Celia Fiennes (1662–1741), who between 1685 and 1703 rode on horseback through the English counties from Northumberland to Lands End; another was Daniel Defoe (1661–1731) who in 1722, not long after he had published 'Robinson Crusoe,' set out on the first of his seventeen "tedious and vastly expensive tours." Forty years later the young French metallurgist Gabriel Jars (1732–69) spent a year visiting British mines, ironworks, factories and the like, his reports on which provide the best view of industrial England at that time.

The object of Defoe was to see how "the fate of things gives a new face to things; produces changes in low life and innumerable incidents; plants and supplants families; raises and sinks towns; removes manufactures and trades; . . . great rivers and good harbours dry up and grow useless; again new ports are opened; brooks are made rivers; small rivers navigable pools and harbours are made where there were none before, and the like." Had Defoe been able to repeat his journeys in 1822 he would have seen plenty of examples of the fate of things giving a new face to things. His first tour took him through the eastern counties and he arrived at Cambridge when the now almost forgotten Stourbridge Fair was being held on what is today Stourbridge Common. Here indeed was a "Britain Can Make It" Exhibition. In Defoe's view the Fair was bigger than those at Leipzig, Frankfort, Nuremberg or Augsburg. Every possible kind of manufacture and ware was there and in abundance, from north, south, east and west. Fifty hackney carriages had been sent from London to ply between Cambridge and the Fair, and some of London's hundreds of wherries had been carried by waggon from the Thames to the Cam. On the river were the barges by which much of the merchandise had been conveyed from King's Lynn, to which place it had been shipped from many a river estuary. Our streams were the channels for the carriage of heavy and bulky goods and one of the great achievements of the eighteenth century was the improvement of inland water transport and of the terminal ports.

The leisurely and all embracing plans of Defoe had no place in the programme of Gabriel Jars. When the Seven Years War was over the French Government sent him to report on our industries, of which they had heard much. Though but half the age of Defoe he was admirably fitted for his task for he had passed through the École des Ponts et Chaussées, and had learnt a great deal about mining, having made several tours to the mines and ironworks of Saxony, Bohemia, Austria, the Tyrol, Styria and Carinthia. He arrived in England in July 1764, and stayed till September 1765, visiting collieries, blast furnaces, foundries, file and saw factories, lead mines, paper works, chemical works, salt works and porcelain factories, taking in his stride Carron, Edinburgh, Glasgow, Liverpool, Manchester, Sheffield, Birmingham, and so down to Cornwall. His six original reports are preserved in the French Archives Nationales, and these M. Jean Chevalier recently said " indicate all the most arresting features which British industry could show in 1765, to a French engineer who had become an extremely well-informed metallurgist as a result of his three years study of the mines and iron works of Central Europe."

It has often been pointed out that at the beginning of the eighteenth century England, in common with other countries of Europe, was largely an agricultural country. So she remained in spite of ironworks and cotton mills, but she exhibited at the end of the century an agriculture vastly improved through the efforts of Jethro Tull (1674–1741), Viscount Townshend (1674–1738), Robert Bakewell (1725–95), Coke of Holkham (1752–1842), and Arthur Young (1741–1820). Young was an unsuccessful farmer, it is true, but he was an indefatigable traveller, an acute observer, and a prolific writer. On the foundation of the Government sponsored Board of Agriculture in 1793, with Sir John Sinclair (1754–1835) as its president, Young became the secretary. With improvements in tillage, cropping, manuring and grazing, came the first of a long series of improvements in agricultural implements, and the invention by Andrew Meikle and others of that great labour-saver, the threshing machine. Agriculture, however, was one of the last industries to be affected much by the inventors and engineers of the eighteenth century, whose work it is the object of this article to review.

II.

Steam Engineering.

If ever there was an invention which satisfied an urgent need it was that of the steam engine as introduced by Thomas Newcomen (1663–1729) of Dartmouth, and improved by James Watt (1736–1819). It is true that Newcomen's was but an " atmospheric " engine in which steam played second fiddle to the pressure of the atmosphere, but it was nevertheless the first heat engine to be used on any considerable scale. Moreover, it was the first notable example of the application of scientific discovery to industry linking the work done in Italy, Germany, the Netherlands and France by Galileo, Torricelli, von Guericke, Huygens, Papin and others with the dewatering of British mines, and this as the result of the genius of an English tradesman.

In the early days of Newcomen power was obtained from horses, water and wind. There were water-wheels on nearly every stream, and windmills formed an arresting feature in the landscape of this and other countries, especially Holland. At one time 500 could be counted in Kent and 2,000 in the Fen districts. The millwright—the predecessor of the mechanical engineer of the present day—worked mainly in wood, but was skilled in the use of the forge and the lathe as well as the axe and the hammer. The windmill as applied to grinding corn or to pumping represented the acme of his art. When neither wind nor water was available, horses provided the power. Some collieries maintained scores of horses for working their pumps, and as late as the end of the eighteenth century the dry docks in the Royal Dockyards were emptied by means of chain pumps driven through horse gears. If it was desired to see the greatest example of millwrighting of the time of Newcomen, it was necessary to visit France and inspect the "Machine de Marli" which raised water from the Seine to the dry plateau 500 ft. above it, on which Versailles is built. This huge machine was constructed for Louis XIV by the Belgian mechanician Swalm Ranniquin, or Renkin, during the years 1675–82. It comprised 14 water-wheels, 253 separate pumps, and literally miles of iron transmission rods, the noise of which when working was "like that of a number of waggons loaded with bar iron running down hill with axles never greased." Like the Versailles palaces it was just one of the extravagances of the king, and was paid for by money wrung from his long-suffering subjects.

In this country it was the much more useful and prosaic object of draining mines that stimulated the inventors of pumping plant, and especially of pumps worked by steam power. Many were the individuals who applied themselves to the problem which was incompletely solved by Captain Thomas Savery (1650–1715) in 1698, and completely by Newcomen almost contemporaneously. Savery's ingenious apparatus was used on a very limited scale; Newcomen's engine, however, came into use in 1712, and atmospheric engines, not differing in principle but only in size, were working two hundred years later. Of Newcomen we know little, and still less of how he came to invent his engine, for neither sketch, plan, calculation nor model of his has survived. His engine included an overhead wooden beam or lever supported at the middle. From one end hung the heavy pump rods; from the other hung a piston, 2 ft. or so in diameter, which worked in an open-topped cylinder of brass. Beneath the cylinder was a boiler of copper such as brewers used fired at atmospheric pressure. When not in use the beam rested with the pump rods end down. The object of the engine was to raise the rods; when released their weight did the rest. Three operations were necessary to start and keep the engine in motion. Firstly, steam had to be admitted to the cylinder; secondly, a spray of cold water had to be injected to condense the steam and thus produce a partial vacuum; thirdly, the water of condensation had to be drained away.

On the formation of the partial vacuum, the pressure of the atmosphere forced the piston to the bottom of the cylinder and the pump rods were thereby raised. In its early form the necessary valves or cocks on the cylinder were probably worked by hand, but all extant illustrations show the engine to be fully automatic. Here was something entirely new to the world, an engine which would work wherever there was fuel to burn and water to boil. Thermodynamically the engine was extremely inefficient, but where coal was cheap this mattered little. The engine could exert more power than horses could be concentrated to do and that sufficed. Known generally as " fire engines " many were erected in Newcomen's lifetime, and by the time of Gabriel Jars' visit in 1764 it is known that about a hundred were installed, one of them with a cylinder of over 6 ft. diameter. Such engines had been introduced by English engineers into various European countries and just when Jars was setting out on his tours of inspection here, Ivan Ivanovitch Pulzunow, a Russian mechanic, was constructing a Newcomen engine at Barnaoul in the mining district of western Siberia. With the passage of time improvements were made in these fire engines; brass cylinders gave place to those of cast iron, and boilers were constructed of wrought iron. The year 1769 was memorable because Smeaton, as the result of experiments with a model, placed the design of such engines on a sound basis. The engines Smeaton was responsible for were among the largest and best of their kind. Though it would be incorrect to say that Newcomen's name had ever been forgotten, it was, however, not till 1921 that any memorial to him was erected. A year before this was unveiled at Dartmouth, his name had been given to the Newcomen Society for the Study of the History of Engineering and Technology, the 'Transactions' of which now form the most valuable corpus of information available to the student on those subjects.

Of James Watt, usually acclaimed as the inventor of the steam engine, but more correctly as its greatest improver, everyone has read. There are more books about him, more memorials to him, than to any other engineer. His correspondence, his sketches, his drawings, his models are all familiar. Of his upbringing and his environment, his ill-health, fits of despondency, want of self-confidence and his inability to deal with men, as well as his versatility and marvellous fertility of invention much has been written. Like his senior, Smeaton, he began his career as an instrument maker in London, during the year 1755–56. There was probably no lonelier youth in the city but assuredly no more assiduous pupil to the trade. Back in Glasgow he became instrument maker to that University, and as such enjoyed intercourse with mathematicians and philosophers Simson, Robison, Adam Smith and Black. The University owned a model of an atmospheric engine, but though it had been overhauled by one of the best instrument makers of the day it would not work. In 1763 it was put into Watt's hands. This proved to be the turning point in his career, and it is probable that for the next thirty years he seldom went to bed without thinking of steam engines. After prolonged thought, calculation and experiment, in 1765 came the flash of inspiration which led to the application to the engine of a separate condenser and air pump, the greatest single improvement ever made. A patent for this, and trials with a full-scale engine seemed out of the question. First Watt borrowed money from Black; then Roebuck, busy at the recently founded Carron Ironworks, took over Watt's debt to Black for a share of the patent which was secured in June 1769. In secrecy at Roebuck's residence at Kinneil an engine was constructed, but with indifferent success. Then Roebuck got into financial difficulties, but the buoyant businesslike Birmingham manufacturer, Matthew Boulton, came to Watt's rescue. To Birmingham went both Watt and the Kinneil engine; in 1775 success was achieved, an extension of the patent for twenty-five years was secured and the famous partnership of Boulton with Watt begun. Soon "Beelzebub," first of Watt's engines, was completed; in 1776 a Watt engine was supplied to John Wilkinson for working a blast furnace blowing engine at Broseley, and that year when Boswell visited Soho, Boulton, on showing him round, exclaimed, "I sell here, Sir, what all the world desires to have. Power!" Much less extravagant in coal consumption than the older "fire" engines, the Watt pumping engines were soon ousting the Newcomen engines, especially in Cornwall where coal was dear; hence by 1780 there were no fewer than twenty at work in that county. What the world needed now was an engine for turning mill work, for by 1781 Boulton was describing London and Manchester "steam mill mad." To the seventeen-seventies belongs the early development of the Watt pumping engine; to the seventeen-eighties the early development of the Watt rotative engine. Inventions flowed one after the other from Watt's prolific brain. Substitutes for the crank, methods of working steam expansively, double-acting cylinders, the parallel motion, the centrifugal governor, the throttle valve, indicators and gauges, etc., all belong to this period. The works at Soho became the centre of the mechanical world, and a great engine factory, Soho Foundry, came into being. This was not only the home of engine-making, but a training ground and a technical school for the rising generation of mechanical engineers. By the end of the century, when Watt retired, some five hundred engines had been constructed.

When Watt was so absorbed with the inventions of the seventeen-eighties there were many engineers experienced in the erection and operation of engines who thought themselves capable of adding their contributions to steam engineering. These men often found their path blocked by the unduly prolonged patent of 1769; yet they pushed ahead. One of these engineers was Jonathan Hornblower (1753–1815), who in 1781 secured a patent for an engine in which the steam was used successively in two cylinders, the forerunner of the later compound engine. Hornblower erected several engines in the West Country, but they proved in no way superior in economy to the Watt engine, and after litigation his clients had to pay royalties to Watt for using his condenser patent. Other men were thinking of the engine as a means for driving boats and vehicles, and others were dreaming of high-pressure engines. An uncle of Hornblower's, Josiah Hornblower (1729–1809), in 1753 had erected a Newcomen engine in New Jersey, North America, but nothing more was done there till just after the American War, when John Fitch (1743–98) and James Rumsey (1743–92) made experiments with steam boats, the one on the Delaware and the other on the Potomac River. There were also contemporary experiments in France by the Marquis de Jouffroy (1751–1832), and in 1788 the Scottish engineer, William Symington (1764–1831), in collaboration with the Edinburgh banker, Patrick Miller, constructed the marine steam engine preserved in the Science Museum, South Kensington. Fourteen years later Symington made a far superior engine for the tug *Charlotte Dundas*, which in 1802 towed barges on the Forth and Clyde Canal. No one has a better claim than he to be called the "father" of marine engineering. As early as 1769, the French military engineer Nicolas-Joseph Cugnot (1725–1804) had constructed the first of all mechanically driven vehicles, and in 1784 William Murdock (1754–1839), Boulton and Watt's most valued assistant, built his model steam carriage. Much of the post-Watt success of the Soho Foundary was due to Murdock. Through him it was the first factory in the world to be lighted by gas, and the first in which compressed air was used for the transmission of power.

By the time of Watt's retirement the stage was set and the actors ready for the next scene in the drama of early mechanical engineering, and the nineteenth century saw engine factories spring up on the banks of the Thames, the Clyde and the Tyne, in Liverpool, Leeds, Hayle and elsewhere. Some of the works then established exist today. If it were possible to carry this review further it would introduce the reader to the work, among others, of John Hall (1764–1836) of Dartford, Matthew Murray (1765–1826) of Leeds, Arthur Woolf (1776–1837) of Camborne, Richard Trevithick (1771–1833), by many regarded as the greatest mechanical genius of his age, William Fawcett (1771–1845) of Liverpool, and Henry Maudslay (1771–1831) of Lambeth.

III.

Iron and Steel Manufacture.

Though the introduction of steam power must ever render the eighteenth century memorable in the annals of material civilization, scarcely less far-reaching in their effects were the contemporary advances made in this country in the manufacture of iron for castings and forgings and of steel for tools, all so essential for the new mechanical engineering. The histories of industries are inextricably interwoven, the demands of one leading to advances in another and these in turn proving of benefit to still others and creating new ones. It was the needs of the collieries and mines which led to the introduction of the steam engine, and this by its capacity and adaptability assisted both miners and ironworkers. The iron industry in its turn made further demands on the miners and engineers, the result being that coal production, iron output and machinery manufacture increased simultaneously.

At the beginning of the eighteenth century when all iron was smelted with charcoal as fuel, and when the production of bar or wrought iron was almost the sole aim of the ironmaster, output was small and iron was expensive. By the end of the century the production of iron had multiplied between thirty and forty times, coal output had increased severalfold, and steam engines were installed at nearly all collieries and ironworks and in many factories. Moreover, the small furnaces and forges situated in the well-wooded counties of England and Wales had been superseded by large establishments situated on the coal fields of the north of England, the Midlands, Scotland, and South Wales.

Outstanding among the many individuals who played their part in this great transformation were the three Darbys, Abraham I (1677–1717), Abraham II (1711–63), and Abraham III (1750–91), whose work was done in Shropshire; the three Wilkinsons, Isaac (d. 1784), John (1728–1808) and William (1738–1808), whose principal blast furnaces and foundries were in Denbighshire, Staffordshire, and Shropshire; Dr. John Roebuck (1718–94), the founder of the Carron ironworks, Stirlingshire; Henry Cort (1740–1800), whose experiments which revolutionized the production of wrought iron were made in Hampshire; and Benjamin Huntsman (1704–76), the founder of the crucible cast steel industry of Sheffield. It must not be overlooked, of course, that other countries contributed to the advancement of the iron and steel industries, or that during the whole century the best wrought iron came from Sweden, always to the fore in mining and metallurgy, or from Russia where Peter the Great early in the century had founded a mining academy and a mining school on the model of that in Saxony.

An ironworks at the beginning of the century consisted perhaps of one small blast furnace for the production of pig iron and a forge with a finery or chafery in which the pig iron was converted into bar iron. These works were situated in isolated spots where iron ore, timber and water power were available. One of such places was Coalbrookdale, some miles south of Shrewsbury, the stream down the valley of which falls into the river Severn. It is " a very romantic spot, it is in a winding glen between two immense hills, which break into various forms, and all thickly covered with wood, forming the most beautiful sheets of hanging wood. Indeed too beautiful to be much in unison with that variety of horrors art has spread at the bottom; the noise of the forges, mills etc. with all their vast machinery, the flames bursting from the furnaces with the burning of coal, and the smoak of the lime kilns, are altogether sublime . . . " So wrote Arthur Young in 1776. To this romantic spot sixty-eight years earlier had come the Quaker brassfounder of Bristol, Abraham Darby, to lease one small furnace there and to start his patented process for making bellied pots in cast iron which should be cheaper than brass pots, and the material for which he succeeded in producing by smelting iron with coke. Unfortunately death overtook him in 1717, when he was only forty years of age and his son but six. The business was then formed into a company and managed for the time by his widow and her son-in-law Richard Ford (1717–45). On coming of age Abraham Darby II. became the head, and under him the works flourished. Smelting with coke was perfected and new furnaces and forges were set up at Horsehay and Ketley to the north of Coalbrookdale. Already in 1724 the firm was using cast iron largely in the production of castings of all kinds such as "cillinders," "pisterns," pump barrels and pipes for "fire engines," and about 1740 a "fire engine" was installed for pumping back water to drive the water-wheel when the stream was low. In 1763 death again intervened, Abraham Darby II. dying at the age of fifty-three, and this time his son-in-law, Richard Reynolds (1735–1816) assumed charge. To him was due the introduction about 1767 of the use of cast iron plates on the wearing surfaces of the many miles of wooden tramways connecting the various works. In 1766 Reynolds assisted two of the leading men of the works, Thomas and Robert Cranage, to secure a patent for "making pig-iron or cast-iron malleable in a reverberatory furnace with pit-coal only." The invention was not brought to a successful issue or it would have anticipated Cort's invention of eighteen years later.

The progressive spirit of the third Abraham Darby, who became the head of the business in 1768, was shown in many ways, and by 1785 the company besides working many iron mines, limestone and coal pits, had ten blast furnaces, nine forges and sixteen steam engines. His most remarkable undertaking was the construction of the cast iron bridge over the Severn between Madeley and Broseley, which still stands and has been scheduled as an ancient monument. Its semi-circular arch has a span of 100 ft., the main members being five ribs with a cross-section of 9 in. by 6 in., each rib being cast in two lengths. The weight of the whole structure is 378 tons. It was by far the largest and finest cast iron structure the world had ever seen, and paved the way for other notable bridges of the same material. Coalbrookdale thus will always be remembered as the birthplace of the use of coke for smelting, the cast iron cylinder, the iron tramroad and the cast iron bridge.

It was but natural that the new method of smelting with coke should gain acceptance first in the Midlands, and it was in this district John Wilkinson became famous. His reputation was perhaps partly due to his association with Abraham Darby III. on the one hand, and with Boulton and Watt on the other. He was, however, the architect of his own fortunes. His father Isaac is first heard of in Cumberland farming and charcoal iron-making. In 1738 when John was ten, the family removed to North Lancashire where Isaac started iron-founding. The same year he took out the first of his four foundry patents, that for a box smoothing iron. In 1753 he migrated again, this time to Bersham, near Wrexham,

Denbighshire, John having meanwhile established himself as an ironmaster at the Bradley furnace, near Bilston, Staffordshire, adopting the methods of the Darbys. After being assisted for some years at Bersham by John, Isaac removed to Bristol, and about 1762 John, then thirty-four, with his brother William, founded the New Bersham Company, the partnership lasting till 1795 when the works were closed. To add to his business John also founded works at Broseley on the Severn, not far from Coalbrookdale. In 1774, the memorable year Watt removed to Birmingham, he took out a patent for an improved boring machine, the accuracy of which appealed so greatly to Boulton and Watt that from 1775 onwards for the next twenty years he had the monopoly of casting and boring cylinders for the firm. In 1776 he obtained from Soho an engine for driving a blowing cylinder for his blast furnaces, thus obtaining the higher air pressure which removed the final objection to smelting with coke. Ever ready to stimulate the iron trade, he assisted Darby in the erection of the cast iron bridge already mentioned, and in 1787 he launched on the Severn the first of several barges or lighters having a sheet iron skin, and thus became the father of the iron ship. His brother William was equally busy though in other ways. After obtaining contracts for the cast iron pipes for the water supply of Paris he established the ironworks at Le Creusot, where he installed blast furnaces, kilns, a foundry, a cannon boring mill, and introduced the use of both cast iron rails and the cupola. He afterwards helped to re-organize the iron industry of Silesia. There was little in common in the outlook of the Wilkinsons and the Darbys, save in their views on the importance of iron, for John certainly possessed nothing of the Quaker spirit.

It is some measure of the influence of the pioneering work of the Darbys that Dr. John Roebuck and his partners, who founded the Carron ironworks, near Falkirk, Scotland, should from the first have used coke and not charcoal. The works, laid out entirely for the production of cast iron, were built in 1759, and on January 1, 1760, with ceremony befitting the launching of a ship, the first furnace was blown in and the casting of pig iron begun. Roebuck, then forty-two years of age, had qualified at Leyden as a physician, but becoming interested in industrial chemistry, about 1749 had set up as a sulphuric acid manufacturer at Prestonpans, using lead chambers instead of glass bells. From chemicals he turned to iron, and by his work at Carron introduced the large scale making of iron into Scotland. When Gabriel Jars visited the Carron works he found there two blast furnaces 30 feet high yielding about two tons of iron every twelve hours, the iron being of excellent quality. Here as elsewhere a difficulty was the insufficiency of the blast and Jars noted that a bellows of prodigious size, 22 ft. in length and incorporating pieces of oak 10 in. thick, was being made. Some years later Smeaton designed a blowing engine for the works. The manufacture of castings of all kinds was undertaken, and as time went on the firm became famous for fire grates, railings, balustrades, urns, panels, plaques, etc., many of which were decorated with designs by architects and artists. From Carron too came the naval "carronade" sometimes known as "the smasher," and early in the nineteenth century the company made experiments with and supplied large numbers of the spherical case shots invented by Colonel Henry Shrapnel.

All this progress in the making of cast iron had little effect on the production of the sister material, wrought iron, and the country remained largely dependent on Sweden and Russia for the best material for anchors, chains, tyres, boiler plates, and smiths' work generally. This situation, however, was completely changed through the work, not of an ironmaster, but of the Navy agent Henry Cort. Through marriage he had become related to one Attwick who had a contract for supplying Portsmouth Dockyard with iron naval stores. In 1772 the contract was assigned to a Mr. Morgan who had a forge at Fontley, two miles from Fareham, Hampshire. The work done there was the reforging of old scrap iron from the dockyard. To Morgan Cort advanced money, but the former's business languishing he took over control, and it was at Fontley he made the experiments which led to his two notable patents of 1783 and 1784, describing (1) the use of rolls with grooves and collars for the production of bar iron, and (2) a method of producing wrought from cast iron in a reverberatory furnace, heated with coal, a process subsequently known as "dry puddling." Cort's inventiveness consisted in the combination of existing techniques to establish new ones, and his methods brought about a revolution in the wrought iron trade comparable to the earlier one in the cast iron trade, and except for iron of special qualities the country became independent of Sweden. The new technique spread rapidly and Cort should have realized a fortune as great as that of John Wilkinson, had it not been that through the death of his partner Adam Jellicoe, who had, it turned out, defrauded the Navy of a large sum, Cort lost everything and he died poor. Before his methods were superseded in the nineteenth century by the inventions of Bessemer, Siemens, Martin and Thomas there were more than eight thousand puddling furnaces in the country.

Into the history and working of blast furnaces, fineries, chaferies, air furnaces, reverberatory furnaces and cupolas, all in use in the eighteenth century, there is no need to enter. The iron industry from primitive times to the nineteenth century is the subject of many scholarly and critical papers by Jenkins, Ashton, Hulme, Hall, Dickinson and others in the 'Transactions' of the Newcomen Society. Neither is it intended to trace the development of the process by which the pure wrought iron of Sweden is converted by cementation into blister steel, from which by "faggoting" is produced the shear steel used for many of the tools sold by ironmongers. These processes were carried on by many manufacturers in or near Sheffield at the beginning of the eighteenth century, and Jars found each of them "ceaselessly occupied in inventing new means of cutting down labour costs and thereby increasing his profit."

Had Jars by chance gained admittance to a small works at Handsworth close by, he would have found Benjamin Huntsman, behind closed doors, trying, not to reduce costs, but to make a steel superior to any Sheffield could produce. A Lincolnshire man by birth, and a Quaker by persuasion, as a watchmaker at Doncaster he was dissatisfied with the steel available for springs, and experimented with a view to finding something better. In 1740 he removed to Handsworth, and it was there he found what he sought by smelting piece of shear steel in crucibles heated by burning coke and casting it in the form of ingots. It took some time to overcome the conservatism of steel users, but by the end of the century Sheffield was producing the finest razors and pocket knives in the world. The works Huntsman founded at Handsworth in 1751 were moved in 1770 to Attercliffe to the north of Sheffield, where the firm of B. Huntsman, Ltd. still makes steel by the founder's method.

IV.

Textile and Other Industries.

The tide of invention which set in with the eighteenth century affected all trades and occupations, as can be seen by the patent records, which show that the number of patents issued in the four successive quarters of the century were 116, 176, 458 and 1349. These included master patents such as those of Watt and Arkwright taken out in 1769, but also minor patents such as that for making blackcurrant lozenges for curing sore throats. But patents cost over £100, and many improvements are not found recorded in any Patent Office. The increasing interest in invention was due to many causes, one of which was the work of the Royal Society. As early as 1663 it was laid down that the business of the Society included the improvement of all useful "Arts, Manufactures, Mechanick practices, Engines and Inventions by Experiments." Gunnery, shipping, the diving bell, Papin's digester and Savery's pump aroused almost as much interest as did air pumps, telescopes and microscopes. In a pamphlet issued in 1718 during the presidency of Newton, giving a list of subjects suitable for contributions to the Society, was an item referring to "New Inventions or Improvements in Mechanicks with descriptions of Machines, Engines, Instruments and the like . . ." In 1738 the Society awarded the Copley Medal to James Vauloue for his pile-driver used in the construction of old Westminster bridge. In 1746 it was given to Benjamin Robins for his experiments on projectiles and the invention of the ballistic pendulum. Three years later John Harrison received it for his work on chronometers, and in 1759 it was awarded to Smeaton for his "Curious experiments concerning Water-wheels and Windmill-Sails." No other engineer received the Copley Medal till it was awarded to Sir Charles Parsons one hundred and sixty-nine years later. What the Royal Society did in the first half of the eighteenth century was followed up by the Society of Arts, ever ready to encourage useful invention. The work of the British clock and instrument makers in the eighteenth century was outstanding, and the names of Tompion, Graham,

Harrison, Mudge, Ramsden and Troughton are familiar to all. There are no finer examples of their art than the chronometers of Harrison and the dividing engine and theodolites of Ramsden, some of whose practices were adopted by mechanical engineers. Contemporary with their work came the invention of stereotyping by Ged and its development later by Alexander Tilloch, the founder of the Philosophical Magazine, the invention of the compound-lever weighing machine by John Wyatt, of the hydraulic ram by Whitehurst, and of the hydraulic press by Bramah. The circular saw of Miller, the screw-cutting lathe of Maudslay, and the woodworking machines of Marc Isambard Brunel and Sir Samuel Bentham, like the boring machines of Smeaton and Wilkinson, all belong to the eighteenth century. Invention appealed to men of all degrees. The ingenious Irish landlord Richard Lovell Edgeworth contrived and worked a telegraph, devised a track-laying vehicle, a dynamometer for ploughs, and wrote on roads and carriages. His contemporary, Charles, third Earl of Stanhope, was equally versatile, making experiments with fire-proof buildings, lime kilns, calculating machines and steam boats, and inventing the Stanhope printing press. Perhaps in this list of benefactors of the arts should be included the pioneers of our ceramic industry, of whom Josiah Wedgwood was recognized not only as a great master potter, an improver of his art, but also as a promoter of canal construction, and " a pioneer and populariser of Watt's steam engine in its application to industry installing an engine at Etruria as early as 1782."

No branch of manufacture benefitted more from invention than the textile industry, which in a hundred years was transformed from a purely domestic industry carried on in homes in towns and country by means of the centuries-old distaff, spinning wheel and hand-loom into a large-scale factory industry with machines driven by water or steam power. Defoe, in his account of the Stourbridge Fair, wrote of the prodigious trade in woollen goods which came to Cambridgeshire from places as far apart as Norwich, Exeter and Halifax, and of the display of cloths, kerseys, pennistons and fustians from Lancashire and Yorkshire. A good deal of technical history centres on these fustians, goods made with warp of flax and weft of cotton. The advent of the new fibre, cotton, had a marked effect on the textile trade and its spinning and weaving had their home in Lancashire, this being partly due to the humid climate of the district. The manufacture of cotton goods met with considerable opposition from the woollen and silk trades, and early in the eighteenth century legislation was passed prohibiting the living from wearing certain cotton goods and the dead from being buried in them. Fustians, however, were in a different category, and by the Manchester Act of 1735 it was definitely laid down that they were exempt from the restriction. In the first Manchester directory, issued in 1772, the names of no fewer than 132 fustian manufacturers were given. Bolton was another centre of the fustian industry. It was easier to spin cotton yarn suitable for weft than for warp, but by the inventions of Hargreaves, Arkwright and Crompton all difficulties were removed.

Looking back it is not a little surprising that spinning by machines came so late, for as early as 1589 the Rev. William Lee (d. 1610) had invented the stocking frame which could do the work of a dozen hand knitters, while the Italians had long possessed machines for doubling and twisting silk. Copies of their machines were to be found in the silk mill of Sir Thomas Lombe (1683–1739), erected in 1718 at Derby. This factory employed 300 workpeople, and its machines for drawing, spinning and twisted silk included 97,746 wheels, movements and individual parts all driven by a water-wheel. The doubling and twisting of silk filaments, however, is fundamentally different from the drawing out and building up of cotton fibres into a yarn, yet this exemplification of factory production preceded by half a century the cotton spinning factory of Arkwright.

Lombe's patent, granted in 1718, expired in 1732, when the silk manufacture was thrown open to all, with benefit to the trade, and it was at this juncture of affairs that the weaving industry received its first notable improvement by the invention in 1733 of the " fly shuttle " for looms, an invention due to the unfortunate John Kay, of Bury, not to be confused with the John Kay of Warrington who afterwards assisted Arkwright. From thence onward invention followed invention and by the end of the eighteenth century the revolution in the technique of the cotton industry had been accomplished. It is impossible in a short space to give an account of the progress in detail, but it may be of interest to set down in chronological order the names of the principal inventors and what improvements they introduced :—

1733. John Kay, of Bury, patented the " fly shuttle " for looms, more than doubling the speed of weaving.

1738. Lewis Paul (d. 1759) patented a machine for spinning wool incorporating principle of drafting by rollers, the technique of which was successfully achieved by Arkwright in his " water frame." Paul was assisted by John Wyatt (1700–66).

1748. Paul patented a carding engine, the technique of which was likewise adopted by Arkwright.

1758. Jedediah Strutt (1726–97) of Derby patented ribbed stocking frame.

1760. Robert Kay, son of John Kay, patented " drop box " for looms, enabling two or more shuttles to be used successively on the same fabric.

1764. James Hargreaves (1720 ?–1778) of Blackburn invented handworked " spinning jenny," an intermittent multiple drafter, twister and winder on. Not patented till 1770.

1769. Richard Arkwright (1732–92) of Preston, with assistance of John Kay of Warrington, patented his " water frame " incorporating improved rollers for drafting, and adapting water power to drive it ; the first power driven spinning machine.

1774–79. Samuel Crompton (1753–1827) of Bolton perfected hand-driven multiple-spindle spinning " mule " capable of spinning the finest counts of yarn. Not patented.

1785. Rev. Edmund Cartwright (1743–1823) of Goadby Marwood, Leicestershire, patented power driven loom. Other patents secured in 1786, 1787 and 1792.

1793. Eli Whitney (1765–1825) in Georgia, U.S.A., invented hand driven " cotton gin " for removing seeds from the cotton bolls. Patented in 1794.

In this list may be added the French inventor Joseph-Marie Jacquard (1752–1834) who, by bringing together the inventions of Falcon, Bouchon and Vaucanson perfected the loom that goes by his name, capable of weaving figured fabrics of the utmost complexity.

Of the great difficulties under which these men laboured, of the opposition they met with from operatives who believed their livelihood to be threatened, and of their unequal rewards much has been written. Of them all Arkwright, who was knighted in 1786, was the only one to realize a fortune. The first power driven cotton spinning factory was that he set up in 1771 at Cromford, on the river Derwent in Derbyshire, with the assistance of Jedediah Strutt. Arkwright erected others at Belper, and Chorley, Lancashire, and entered into partnership with the philanthropic industrialist David Dale (1739–1806) in establishing a mill at New Lanark, near the Falls of Clyde.

The cumulative result of the inventions and enterprises referred to was to place Great Britain far ahead of any other in the manufacture of cotton goods and to establish at the hands of mechanical engineers a flourishing textile machinery industry. In the story of trade there is no more remarkable chapter than that which traces how a natural fibre grown in hot climates has been brought year after year thousands of miles by sea to the damp valleys of an English county and then after manufacture taken back those thousands of miles for the use of the peoples who have grown the cotton plant. Engineers, however, today manufacture and export air-conditioning plant which enables cotton mills to be operated anywhere.

V.

Progress in Civil Engineering.

A review of the civil engineering in the eighteenth century reveals the same remarkable progress as in industry, the constructors endeavouring to meet the ever-growing demands of the day. There was ample scope for improvement, for the state of the roads, waterways, bridges and harbours of England was far below the standard necessary for rapid communication and cheap transport. The Normans had built great castles, the Plantagenets magnificent cathedrals and the Tudors stately mansions, but all undertakings, classed today as public works, were almost completely neglected. The inhabitants of London had been content for five centuries with the single inadequate bridge over the Thames constructed in 1209, and when in 1710 Wren's son placed the last stone on the summit of the lantern of St. Paul's Cathedral the streets of London were still undrained. Thirty-five years later when the House of Commons discussed a bill for the better paving of Westminster it was said that the streets were a disgrace to the nation and abounded in such filth as a savage would look at with amazement. The country roads were no better than the streets and every traveller, whether he went on foot or horseback, by waggon, post-chaise, stage-coach or the later mail-coaches, grumbled. Thanks to numerous Turnpike Acts there were stretches of road in fairly good condition, and it was on one of these that Johnson was riding in a post-chaise in March 1776 when he remarked to Boswell that " Life has not many better things than this." The experiences of most travellers were more like those of Sydney Smith who wrote that during a nine hours journey from Taunton to Bath he had " suffered between 10,000 and 12,000 severe contusions." Journeys by road were tedious, expensive, uncomfortable and sometimes dangerous, and it was not till the end of the eighteenth century that engineers seriously addressed themselves to a study of road making.

With water transport it was a different story. The rivers of a country are its natural highways, and England was fortunate in possessing numerous navigable streams. Schemes for improving these were often brought forward, and a correspondent of the Chancellor of the Exchequer wrote in 1697 that " they had been greatly plagued in the house with petitions about rivers being made navigable." The House was plagued a good many more times in the eighteenth century, in which canal building began in a small way, grew into a great business and rose to a mania. Side by side with the making of canals, with their embankments, locks, aqueducts and tunnels, came the straightening and deepening of rivers, the building of bridges and the construction of docks and harbours. The state of the Fen district had engaged the attention of both James I and Charles I, and in the eighteenth century nearly every civil engineer was concerned with Fen drainage schemes. With these activities, for the first time came into being a class of men who were wholly engaged in carrying out surveys, drawing up plans, writing reports and supervising constructions, thus ushering in the new profession of civil engineering. The most famous of these early civil engineers included John Perry (1670–1733), James Brindley (1716–72), John Smeaton (1724–92), Robert Mylne (1734–1811), William Jessop (1745–1814), Thomas Telford (1757–1834), and John Rennie (1761–1821). There were others almost equally active, though the biographers have overlooked them. Something of their careers, however, can be gleaned from two papers in the ' Transactions ' of the Newcomen Society, Vols. XVII. and XVIII. (1936–38). The first of these papers was entitled " The Society of Civil Engineers (Smeatonians) " and was by Mr. S. B. Donkin; the second was on " The Early Smeatonians " and was by Mrs. E. C. Wright, Ph.D. The first was based on the records of a society still active, which was founded in 1771 and reconstituted in 1792; Mrs. Wright's paper was the outcome of painstaking research at the British Museum and elsewhere. It is from these studies that information is obtained of the labours of Smeaton's contemporaries, John Grundy, Senr. (d. 1748), John Grundy, Junr. (1719–83), Thomas Yeomans, F.R.S. (d. 1781), John Golbourne (d. 1783), Robert Whitworth, "who levelled for that great Genius Mr. Brindley," Joseph Nickalls, and others. Mrs. Wright pointed out that in the early part of the century some of these individuals called themselves engineers, but that the term "civil" was not added till later on. Most of their work was connected with private undertakings, financed by private individuals, and the engineers were in no way government servants. It was otherwise in France, where the famous Corps des Ponts et Chaussées was founded in 1716, followed by the opening in 1747 of the École des Ponts et Chaussées, where a grounding in scientific principles was given. These institutions boast of a long line of eminent engineers such as Perronet, Chezy, Gauthey, Lamblardie, Prony and Girard, who were responsible for the public works of France in the eighteenth century.

Of the many hundreds of undertakings the British civil engineers mentioned were concerned with, only a very few can be referred to, but some conception of the advances made can be gleaned from brief notices to which attention must be confined, of one or two works carried out at the beginning, in the middle and towards the end of the eighteenth century.

Two early works on a considerable scale were the construction of the first dock at Liverpool by Thomas Steers from which that city's great dock system has been developed, and the repair of the breach in the banks of the Thames at Dagenham by John Perry, after the unsuccessful attempt by another "undertaker," William Boswell. None of the eighteenth century civil engineers was trained as such and Perry was no exception. He started life as a naval officer, but having fought an unsuccessful action against French privateers was court-martialled, fined £1,000 and committed to the Marshalsea for ten years. This was in 1693. Five years later he secured his release and entered the service of Peter the Great, becoming Comptroller of Maritime Works. He spent three years on the cutting of a canal from the Volga to the Don, another eight years in the Voronezh district constructing docks and improving the navigation of the River Voronezh, and another period surveying routes for canals to join Lake Ladoga with the Neva. He returned to England in 1714 and related his experiences in his entertaining book 'The State of Russia under the Present Czar' (1716). The great breach of the banks of the Thames of 1707 had flooded several thousands of acres of land, and a contract was given to Boswell in 1714 for the necessary repairs. The task proving too much for him it was given to Perry. The work took five years and cost about £25,000. Perry was also known for his proposals for improving the harbours of Dover and Dublin, and at the time of his death at Spalding he was engineer to the Adventurers of Deeping Fen, Lincolnshire, a post he was succeeded in by John Grundy, Senr.

The most considerable undertaking after Perry's repair of the Dagenham breach was the building of the old Westminster bridge over the Thames by the French engineer Charles Labelye (1705–81 ?) during the period 1738–50. This was followed a decade later by the building of the Blackfriars bridge to the designs of Mylne. Both bridges were of stone, but whereas Labelye retained the use of circular arches, Mylne adopted elliptical arches over which there was much controversy. The two bridges remained in use for about a century, by which time three others had been thrown across the Thames. In between the building of the Westminster and Blackfriars bridges came the erection during 1756–59 of Smeaton's noble tower on the Eddystone rock. Winstanley's curious lighthouse had been swept away in the great storm of 1703, and John Rudyerd's fine wooden tower was destroyed by fire in 1755 after standing for forty-six years. When consulted about a new lighthouse Smeaton had had no experience as a civil engineer, although he had become an authority on all kinds of mill work ; his work on the lighthouse was therefore the more remarkable. Constructed of stone, with every block dovetailed and secured to all adjacent ones, and with the lower part completely solid, it was completed in 1759 and was a model for other lighthouses in exposed positions. Today the stump still stands on the rock, but the tower has been re-erected on Plymouth Hoe, as fine a memorial as any engineer could wish for. It was not the state of the tower, but the undermining of the rock which led to the building of another one hundred and fourteen years after Smeaton and his friend Dr. Zachariah Mudge had together sung the Old Hundredth in its lantern, open to visitors today.

The building of the Blackfriars bridge coincided with the construction of the Bridgewater Canal by Francis Egerton, the 3rd Duke of Bridgewater (1736–1803), assisted by Brindley whose native talents had gained for him the reputation of one of the most capable millwrights in the Midlands. The first section of the canal connected Worsley with Manchester. This was only seven miles long, but it had to cross the valley of the Irwell, and it was due to the representations of Brindley that the use of locks was abandoned in favour of the Barton Aqueduct, the first of its kind in the country. Begun in 1759, the canal was brought into use in 1761, but its extension to Runcorn on the Mersey took more than ten years, and by the time of its completion Brindley had passed away. In those ten years or so he had acted as engineer for many other canals including the Great Trunk Canal connecting the Mersey with the Trent. To the Duke of Bridgewater canals brought great wealth, to Brindley fame, and to the country an urgently needed means of transport for its coal, iron, manufactures and agricultural produce.

When Smeaton died in 1792 leaving many monuments to his genius, such as the Perth, Banff and Coldstream bridges, Ramsgate Harbour and the Forth and Clyde Canal, Telford was thirty-five and Rennie but thirty-one, and many of their most important works belong to the first quarter of the nineteenth century. The construction of the Ellesmere Canal by Telford and of the Kennet and Avon Canal by Rennie, however, occupied the closing years of the eighteenth. Both Telford and Rennie were of Scottish birth, Telford beginning his career as a stonemason, and Rennie as a millwright, but whereas Telford was almost entirely self-educated Rennie was fortunate enough to spend three winters at Edinburgh University, being the earliest British engineer to enjoy such an advantage. After working at his trade in Edinburgh Telford was employed on the building of Somerset House, London, and the Commissioner's House at Portsmouth, where his talents and industry soon attracted attention. In 1788 he was appointed Surveyor to the County of Shropshire, and five years later engineer to the Ellesmere Canal. This waterway had to cross the river Ceriog at Chirk and the river Dee at Pont-Cysylltau, and in both instances Telford used aqueducts instead of locks. The larger of the two, the Pont-Cysylltau aqueduct, has a length of 1007 ft., and is carried on 19 arches, the piers of which are 121 ft. high. The most remarkable feature of the work, however, was not its size, but its waterway formed of cast iron girders and plates, all firmly bolted together, the ironwork being supplied by William Hazeldine (1763–1840), who afterwards forged the great chains for Telford's suspension bridge over the Menai Straits. Combining the skill of the architect, the civil engineer, the mechanical engineer and the ironfounder, the Cysylltau aqueduct was described during its construction as " among the boldest efforts of human invention in modern times." It was begun in 1795 and completed ten years later.

Rennie's first work in England was the fitting up of the flour-milling machinery and its gearing at the great Albion Mills erected by Boulton and Watt, and others during 1784–88 on the banks of the Thames near the southern end of Blackfriars bridge. There were no fewer than twenty pairs of millstones, the whole being driven by two Watt engines. In size and capacity the mill far surpassed anything hitherto erected, and Boulton looked upon it as a great advertisement. Unfortunately, in March 1791, it was completely destroyed by fire. It had, however, established Rennie's reputation, and he then set up in business for himself, carrying out both mechanical and civil engineering work. His Kennet and Avon Canal opened in 1799 had no such remarkable features as the Pont-Cysylltau aqueduct, but it was pronounced one of the best executed in the country. With a length of fifty-seven miles it had seventy-nine locks and several aqueducts, and allowed barges to pass for the first time from the Thames to the Avon at Bath, a matter of some strategic importance.

With the opening years of the nineteenth century the operations of both Telford and Rennie extended rapidly, and their works are to be seen in many parts of the United Kingdom. From Shropshire Telford was sent to Scotland, where he built some hundreds of bridges and hundred of miles of roads, and also the Caledonian Canal. His Menai Straits bridge was built during 1819–25, by which time he was the recognized head of his profession, and had become the first president of the Institution of Civil Engineers. Rennie's later works were largely connected with the Navy and the City of London. He was the trusted adviser of the Admiralty, and constructed both Sheerness Dockyard and the Plymouth Breakwater. He also designed the London and East India Docks, and the old Waterloo and Southwark bridges, and the present London bridge, built under the supervision of his son Sir John Rennie. Closely associated with Rennie in many of these undertakings were Sir Edward Banks (1770–1835) and his partner William John Jolliffe (1774–1835), the contracting firm of Joliffe and Banks being the largest of its day. Jolliffe and Banks it may be recalled are buried and commemorated in the neighbouring Surrey parish churches at Merstham and Chipstead, but Rennie, like Wren and Mylne, is buried in the crypt of St. Paul's Cathedral, and Telford in the nave of Westminster Abbey, fit resting places for two great representatives of the eighteenth century civil engineers, " a self-created set of men, whose profession owes its origin, not to power or influence, but to the best of all protection, the encouragement of a great and powerful nation . . . "

SCIENTIFIC INSTRUMENTS IN THE 18th CENTURY

By ROBERT S. WHIPPLE, M.I.E.E., F.Inst.P.

THE 18th century witnessed a great development of experimental work which necessitated the use of scientific instruments. Frequently the experimenter himself designed and made the apparatus ; more frequently, he obtained the co-operation of an instrument maker who, working with him, designed and made the desired equipment. The century produced a galaxy of famous scientific men and also a group of instrument makers of outstanding merit who worthily supported their scientific colleagues.

At the commencement of the century the astrolabe, the backstaff and cross-staff were used for determining the latitude of a ship at sea. These instruments were all of them cumbersome and fairly inaccurate in their results. At the end of the century, Hadley's Octant (1731), which at the suggestion of John Campbell (1757) was enlarged into the sextant so as to measure angles up to 120°, was in universal use. This is not surprising when it is considered that in all but the smoothest seas it must have been impossible to obtain readings of any accuracy with the backstaff. On the other hand, with the octant, the reflected image of the sun remains bisected by the horizon during a large movement of the instrument, so that a good reading of the altitude is possible, even in a fairly heavy sea. The scales of the instruments were generally divided on ivory or boxwood to a third of a degree and were fitted either with diagonal scales, or a vernier, so that readings could be taken to one minute of arc. The instruments were sometimes fitted with low-power telescopes in the place of simple pin-hole sights.

The determination of longitude is a more difficult problem in that the local time at some instant must be compared with the standard time corresponding to that of the prime meridian. The difference between these two times gives the longitude from the prime meridian.

As the Astronomer Royal mentions in his paper " Astronomy through the Eighteenth Century," the British Government in 1713 offered a large award for a method of determining a vessel's longitude at sea to within an accuracy of 30′ at the end of a voyage to the West Indies. This corresponds to an error of two minutes in time at the end of a six weeks' voyage. A chronometer, or marine timekeeper, was designed and made by John Harrison, a Yorkshire carpenter. Between 1728 and 1759 he designed and constructed four instruments and with the fourth obtained the award.

In 1766 a Frenchman, Pierre Le Roy, produced his " montre marine " which was fitted with the first compensation balance and this was almost universally adopted by other makers. Arnold and Earnshaw, both ingenious craftsmen, developed the making of chronometers on a successful and a commercial scale, so that the instrument remained practically unchanged until the end of the 19th century.

The invention of the pendulum clock by Huygens in 1657 and of the anchor escapement by Hooke a few years later made possible a great advance in timekeeping. For astronomical work, however, the variations in timekeeping owing to changes in length of the pendulum with temperature were serious and Graham in 1721 was the first to attempt to overcome the difficulty. He used a steel pendulum rod which carried a bob consisting of a steel tube containing mercury and the quantity of mercury was so adjusted that when heated its upward expansion compensated for the downward expansion of the rod.

A few years later Harrison introduced his "grid-iron" pendulum in which use was made of the differential expansion of brass and steel rods to which the bob was attached. About 1800, zinc and steel were used, as owing to the greater expansibility of zinc, fewer rods were required. Until the invention of invar (an alloy of nickel 35·7 per cent., steel 64·3 per cent.) by C. Guillaume in 1895 the standard observatory clock remained practically unchanged since the middle of the 18th century. At his private observatory at Kew, George III. had a collection of such clocks made by the leading clock-makers of the day. Several of these clocks are still in use at the present time.

The Astronomer Royal, in his article previously mentioned, has dwelt on the evolution of the telescope regarded from the point of view of the astronomer. He points out that the compact size of the telescopes with achromatic objectives made the equatorial form of mounting practicable. For the same reason the invention had a great influence on the development of surveying instruments. This is understandable, because the difference in principle between a telescope mounted to survey the heavens and one used to measure distances upon the earth's surface is a small one.

It must also be remembered that both groups of instruments, astronomical and terrestrial, were made by the same men and that in many details the construction was the same. As in the case of the early astronomical instruments, the surveying instruments of the same period were fitted with pin-hole sights, but about 1715 the better instruments were being fitted with telescopes (unachromatic) provided with cross-wires. Some surveying instruments made about the middle of the century are in existence and are examples of excellent workmanship.

As the demands of the astronomer and the surveyor for accuracy increased, so the need for improved divided circles became more insistent. The accurate dividing of circles has always been one of the more difficult tasks of the instrument maker. For many years there was no alternative but to bisect, or trisect, with a beam compass or other tool, the spaces set out on the scale or circle, and to continue this operation until the scale was completed. The accuracy obtained by some of the makers at this time was extraordinary. It is stated that John Bird (1709–1776) was able to obtain an accuracy of five minutes of arc on his 8-ft quadrant by continued bi-section of the arc.

Henry Hindley of York, about 1739, completed a small machine for cutting the teeth in clock wheels and for dividing instrument circles. In 1766 Jesse Ramsden (1735–1800) made his first circular dividing engine and in 1775 his second one which was the pioneer of the modern dividing machine. It was with this machine that Ramsden divided the circles of the 3-ft. theodolites used in the principal triangulation of Great Britain and Ireland. They were divided to 10 minutes of arc and were read to one second by three micrometer microscopes. One of these instruments is exhibited in the South Kensington Museum.

With the exception that in 1826 William Simms invented a self-acting mechanism by means of which the dividing machine became completely automatic, no fundamental improvements were introduced into the machine for nearly a century.

Mention must be made of the telescope eye-piece invented by Ramsden and universally known by his name. It gives a real, or positive, image and for this reason is generally used in optical measuring machines.

The thermometer, or thermoscope, invented by Galileo early in the 17th Century had become a fairly satisfactory article by the use of two fixed points and the division of the interval between the points into a number of equal parts. It was, however, to Fahrenheit (1686–1736), a scientific instrument maker in Amsterdam, that the modern form of thermometer owes its inception. In 1721 he constructed the first mercury thermometer and with it determined the boiling points of a variety of liquids. He standardized his thermometer by taking as his zero the point he defined as " the greatest cold, " obtained with a mixture of salt and ice, and his highest point, or blood heat, obtained by inserting the thermometer in the mouth or armpit of a human body; this corresponded to 96°. It is not known how he obtained his scale of temperature, but it is thought that he divided the distance between these two points into 24 parts, and later subdivided these into four parts, thus arriving at a scale of 96°. On this scale the freezing point of water was 32° and when he extended the scale to the boiling point of water it became 212°. He does not appear to have used the boiling point as a " fixed point " on his scale.

The French naturalist, Réaumur (1683–1757), unaware of the work of Fahrenheit, published a paper in 1730 in which his " rules for constructing thermometers with comparable graduations " were stated. He used alcohol instead of mercury as his thermometric liquid owing to its higher coefficient of expansion. He stressed the importance of calibrating the thermometer tubes and also the purity of the alcohol employed. On Réaumur's scale the melting point of ice was taken as 0°, and the boiling point of water at standard pressure as 80°.

In 1742 the Swedish astronomer Celsius described his method of making a mercury thermometer. In this the scale had two fixed points, that of melting snow and that of boiling water, the barometric pressure having its mean value at the time of the experiment. The stem between the two points was divided into 100 parts. Celsius placed his zero at the boiling point and the 100° at the freezing point. The centigrade scale in which the values were reversed was introduced by Christin of Lyons in 1743.

The well-known Six's maximum and mimimum thermometer was introduced by James Six in 1782. In the modern form of his thermometer the indices containing a small piece of steel and a fine glass spring are re-set by means of a magnet just as he described in his original paper. The modern meteorologist, for accurate work, still uses the maximum and mimimum thermometers of John Rutherford, invented in 1790. The two thermometers were mounted horizontally. The maximum temperature was shown by a mercury thermometer ; an index, consisting of a conical piece of ivory, rested on the surface of the mercury. When the temperature rose the index was pushed before it, when it fell the index was left behind, thus recording the maximum temperature. In the modern instrument a thread of mercury separated by an air bubble has taken the place of the ivory index. The mimimum temperature was recorded by an alcohol thermometer containing a small glass index completely immersed in the spirit. If the temperature fell, the alcohol contracted, pulling the index down with it. If the temperature rose, the spirit passed over it and the index recorded the mimimum temperature.

The community owes a great debt to De Luc (1727–1817) for the work he did in perfecting the barometer. He was primarily interested in using it for the measurement of altitudes and came to the conclusion that the reason for the discordant results he was obtaining was due to the unreliability of his instruments. About 1750 he found that the chief source of error was due to air and moisture included in the mercury column. He boiled a series of barometer tubes, thus driving out the air and moisture contained in them, and found that he obtained agreement in their readings. He also determined accurately the effect of temperature on the length of the mercury column.

As a result of this work he found that the thermometers he was employing were as unsatisfactory as the barometers. He investigated a number of liquids and decided that mercury was the best thermometric substance. His method of calibrating the thermometer tubes and his method of filling and boiling the mercury in them are identical with those followed by the modern maker. It is not incorrect to state that the technique of thermometer manufacture at the close of the 18th century remained practically unchanged until John Welsh in the middle of the 19th century pointed out the importance of prolonged annealing of the thermometer bulb, and the continental physicists introduced about fifty years later the new hard glasses from which to make the bulbs and stems.

De Luc's long and thorough investigation of the sources of error in the barometer had so much advanced the manufacture and accuracy of this instrument that barometers made by Newman at the commencement of the 19th century are still in use at the Greenwich and Kew Observatories.

A distinguished geologist, De Saussure (1740–1799), published an account of his hair-hygrometer in 1783, an instrument which he had long been developing. A taut hair lengthens when wetted and contracts when dried, over a range which may be about one-fortieth of its length. De Saussure devised simple instruments by means of which the changes in length were caused to move a needle over a dial. He devoted a great deal of time to the selection of his hairs, to freeing them from grease and to an exhaustive study of their performance. All this was the work of an able experimenter and has stood the test of time. The latest application of his hygrometer is in the little meteorograph attached to the balloons used for determining the humidity of the air at high altitudes.

The calorimeter developed by Lavoisier and Laplace and described by them in 1784 was devised for the determination of the specific heat of substances. They defined the unit of heat as the quantity of heat required to raise the temperature of a pound of water by one degree. The specific heat of a substance was expressed as the ratio of the heat required to raise the temperature of a given mass of the substance to the quantity required to raise an equal mass of water through the same range of temperature. In their calorimeter the specific heat was determined by noting how much ice a known quantity of the substance heated to a definite temperature would melt. The quantity of water so obtained was collected and weighed. This mass of water, divided by the product of the mass of the body and the number of degrees of temperature it was originally above zero, is proportional to the specific heat of the body. In its general form the calorimeter remains the same to-day although in detail it has been much elaborated.

John Ellicott's " instrument " (1736) and John Smeaton's " pyrometer " (1754) for the determination of the coefficient of expansion of metals should be mentioned. In Ellicott's instrument the expansion of the metal under test was compared with that of a standard iron bar, and in Smeaton's pyrometer with that of a brass rod, the expansion of which had been determined against that of a bar of wood. In both instruments great ingenuity was shown in designing the details of the measuring mechanism.

In connection with the geodetic survey previously mentioned, Ramsden in 1785 devised a piece of apparatus, generally known as " Ramsden's pyrometer," for measuring the expansion of a metal bar against standard bars maintained at a definite temperature. The observations were required to define accurately the length of a geodetic base line on Hounslow Heath. Three parallel troughs were taken, over 5-ft. long, the outer ones each containing an iron bar kept at a constant temperature by a packing of pounded ice, and the inner one, suitably heated, containing the bar to be tested which was fixed at one end. The change in length of the bar was determined by means of a simple but efficient optical system *.

The development of the telescope and its mountings enormously aided astronomy and geodesy and the concurrent evolution of the microscope had a considerable influence on the biological sciences.

The publication of *The Micrographia* by Robert Hooke (1665) was a great milestone in the development of the microscope. In this book he described a variety of objects he had studied and of which he gave illustrations. The book aroused much interest both in this country and on the Continent and created a demand for microscopes. The introduction of the field lens by Hooke to enlarge the field of view, although originally due to Monconys, was such an improvement that it was generally adopted.

At the commencement of the 18th century the Hooke form of microscope was being made by John Marshall, one of the great English opticians. The first description of his instrument was published in Harris' *Lexicon Technicum*, 1704, and there are in existence several instruments made by him.

The circulation of the blood had been announced by Harvey in 1628 and its demonstration in the tail of a fish or in the web of a frog's foot was of great interest to users of the microscope. Marshall advertised his instrument as "New Invented Double Microscope For Viewing the Circulation of the Blood." Technically the microscope was a great improvement on its predecessors; it was provided with six objectives ranging from a power of four diameters to that of one hundred and was fitted with a screw fine-adjustment for focusing.

About 1738 Lieberkühn (1711–1756) devised a combination of a simple lens, mounted in a polished metal reflector, as a means of illuminating opaque objects. The construction was frequently modified so that the Lieberkühn could be applied to various objectives and it is so used to the present day.

The invention of the solar microscope is generally attributed to Lieberkühn although there is reason to think that it may have been due to Fahrenheit †. In the early instruments a beam of sunlight was thrown directly, into the microscope, by a lens mounted in a fitting placed in an aperture in a shutter, the image of the object was thus projected on to a white screen. About 1743, John Cuff (1708–1772), the inventor of the Cuff microscope, originated the form of solar microscope in which the tube was fixed and the motion of the sun was allowed for by projecting a beam of sunlight, by means of a movable mirror, into the microscope.

---

* I am indebted to Prof. A. Wolf's 'History of Science Technology, and Philosophy in the Eighteenth Century' (George Allen & Unwin, Ltd.) for this reference and also for a good deal of information used in this article.

† See 'The History of the Microscope,' by R. S. Clay and T. H. Court, (Charles Griffin and Company Ltd.), p. 213.

A great many forms of such microscopes were made by various makers and excellent instruments made by Benjamin Martin, George Adams, Ramsden, Dollond and others are still in existence.

Benjamin Martin (1704–1782) was one of the great pioneers in the development of the microscope during the 18th century. He designed and made the first drum microscope in which the mirror was fitted in the cylinder base. This instrument was the forerunner of the cheaper form of microscope so largely made by Continental manufacturers during the 19th century. Although Martin did not claim to have invented the idea of embodying in his microscope the " between-lens " which made his objective into a compound one, thus greatly improving the definition obtained, there appears to be little doubt that he was the first to do so.

He was the first maker to fit a screw-micrometer (1742) in the eye-piece of the microscope so that the magnifying power of the object-glass could be determined. He appears to have suggested the rotating multiple lens-carrier nose-piece in which he placed different powers to rotate successively into the axis of the microscope. His " Grand Universal Microscope " made about 1780 was provided with a mechanical stage having rectangular movements made by fine-threaded screws the heads of which were graduated.

It is not generally known that Martin was the first to construct an achromatic microscope objective. In his *New Elements of Optics*, London, 1759, in the year following Dollond's communication to the Royal Society of his construction of an achromatic telescope objective, Martin dealt with the principle of achromatism and compares the result he obtained with an achromatic object glass and " Speculum, applied as an Object-Glass in a compound Microscope " *. From the comparison he decided that the results obtained from the speculum were much to be preferred to those from the achromatic objective. An interval of about sixty years was to pass before Tully in London and Chevalier in Paris produced satisfactory achromatic microscope objectives.

Martin specialized in making cabinets of optical instruments containing a solar microscope, a drum microscope, a scioptic ball, a large variety of accessories and a telescope. Some of these cabinets are excellent examples of craftsmanship. It should also be mentioned that Martin published a large number of books for popularizing science which had a considerable vogue and which undoubtedly helped in the sale and use of scientific instruments, globes, etc. The two George Adams, father (1708–1773) and son (1750–1795), did a great deal to develop the microscope and its uses. The book by the father, *Micrographia Illustrata*, 1746, shows the great variety of instruments made by the firm, practically covering all the types in general use and it also gives illustrations of many objects suitable for the microscope.

To Adams, senior, also is due the suggestion of screwing two or three

* See ' Cantor Lectures on The Microscope,' by John Mayall, junior, Society of Arts, London, 1885, p. 48.

microscope objectives in series with one another so that the power of the combination could be varied. This method of varying the power of a microscope is still in use in inexpensive instruments. Adams was the first to introduce the method of supporting a microscope on trunnions at its centre of gravity and also developed a micrometer capable of reading to 1/10,000 inch. He made a great variety of instruments for instructional purposes. Many of these for teaching the principles of mechanics were based on J. T. Desagulier's translation of the work of Gravesande, *Mathematical elements of Natural Philosophy. Confirmed by Experiments : Or an Introduction to Sir Isaac Newton's Philosophy*, sixth edition, 1747, and those for teaching Optics on Robert Smith's *Compleat System of Opticks in Four Books*, 1738. Excellent examples of these instruments can be seen in the King's Collection (George III.) now at the South Kensington Museum.

The Adams also specialized in the manufacture of globes. In 1766 Adams senior published the first edition of his book, *A Treatise describing and explaining the construction and use of new celestial and terrestrial globes.* The book had a great vogue, passing through thirty editions.

The French physicist Pierre Bouguer (1698–1758) and his German contemporary Johann H. Lambert (1728–1777) developed two simple photometers and carried out, independently, a series of experiments on photometry, the results of which have remained almost unchallenged till the present day. The photometers have been greatly improved in detail, but the underlying principles have remained.

Great interest was shown in optical instruments and in optical illusions towards the end of the 18th century. Magic lanterns, the illuminant generally consisting of one or more candles, with crude hand-painted slides—frequently of astronomical subjects—were the playthings of the well-to-do. Spy-glasses of exquisite workmanship were to be found in the opticians' shops of London and Paris. Spectacles, as we know from the trade cards of various opticiaus, were slowly coming into use *.

The 18th century saw the birth of the first efficient glass electrical machine by Francis Hauksbee, who described his machine and experiments made with it in his book *Physico-Mechanical Experiments on Various Subjects, etc.*, 1709. Several experimenters developed machines and an excellent example of one made by Edward Nairne (1726–1806) stands in the hall of the Royal Institution.

The invention of the Leyden jar by E. G. von Kleist in 1745 led to an intense interest in electrical experiments. Franklin published a theory of the Leyden jar and his experimental results during the years 1747–1755.

Charles Augustin Coulomb (1736–1806) about 1784 invented his

* The following is an extract from the trade-card of Edward Scarlett (1677–1743) in which it states that he " Grindeth all manner of Optick Glasses, makes Spectacles after a new method, marking the Focus of the Glass upon the Frame, it being approved of by all the Learned in Opticks as ye Exactest way of fitting different Eyes."

torsion-balance which was to become so fruitful in obtaining electrical and magnetic data. He showed that the repulsion between two similarly electrified spheres varies inversely as the square of the distance apart of their centres. By use of practically the same apparatus he determined the law according to which the force of a magnet pole varies with the distance.

Various experimenters had worked at the problem of an electrometer, in which two moving members were mutually repelled from each other by an electric charge, but the first satisfactory one was made by William Henly in 1770. In this instrument a cork ball was hung by a light rod from the centre of a graduated quadrant over which the rod turned. The charge was measured by the deflection of the rod.

Abraham Bennet (1750–1799) appears to have been the inventor of the well-known gold-leaf electrometer, the instrument which in the hands of the workers at the Cavendish Laboratory was to become, in the early days of the 20th century, such a valuable tool.

Alessandro Volta (1745–1827), who had previously invented his electrophorus and condenser, described in 1800 in a letter to Sir Joseph Banks, his invention of the crown of cups—the first battery. The first man in England to make a battery in accordance with Volta's letter was a London surgeon, Sir Anthony Carlisle, and later William Nicholson co-operated with him. It is of particular interest at this time to note that Nicholson described the experiments in *Nicholson's Journal* of July 1800, pp. 179–191, later absorbed by the *Philosophical Magazine.*

There are in museums many cases of drawing instruments made in the 18th century, especially those made by French makers for artillery officers and engineers, which show that the form and construction of many of the instruments have remained practically unchanged till the present day. Some of the ivory and metal scales made during the 18th century are examples of excellent workmanship.

The slide rule was developed primarily as an aid for Custom's officers for determining the volumes of casks, etc. John Robertson (1712–1776) Librarian to the Royal Society, designed a slide-rule 30 in. long and 2 in. wide, having 12 logarithmic scales of natural numbers and trigonometrical functions engraved on one face. He also invented the cursor for use on the rule.

In 1787, William Nicholson (see *Phil. Trans.* 1787, Pt. II. p. 246, and *Nicholson's Journal*, 1797, p. 372) urged the use of concentric circular, or spiral, scales so as to obtain the equivalent of long rules. This has now become standard practice and in the case of the Fuller rule has been developed to a high degree of accuracy.

In a short article it has only been possible to touch upon the development of a few representative instruments. In chemistry, engineering and the biological sciences instruments were made for the solution of special problems and this experimental work has been continuously maintained until the present day.

THE SCIENTIFIC PERIODICAL FROM 1665 TO 1798.

By DOUGLAS McKIE, D.Sc., Ph.D.

BEFORE the foundation of the scientific academies in the middle of the seventeenth century, there were no scientific periodicals ; and " natural philosophers " conveyed their ideas and accounts of their work and experiments to one another by means of letters which, from their scope and length, might be more appropriately described as dissertations. Much of this voluminous correspondence has survived ; and in it is to be found the give-and-take, and sometimes the cut-and-thrust, of scientific argument and debate. We need remind ourselves only of the extensive and historically important correspondence of Mersenne [(1)], from which today we obtain, or shall obtain when it has been completely published and studied, a real, or at least a deeper, knowledge of the history of physical science in the first half of the seventeenth century.

But the epistolary dissertation was not an ideal method for the communication of scientific fact and theory, even when it transcended frontiers. It was too personal. Men write to their friends, and not always, or not so often, to those who dispute their facts and reject their theories. Questions of priority were so easily raised ; and ciphers were used for secrecy. Moreover, the method could not spread new knowledge and new ideas either rapidly or widely : it was too slow and too limited within narrow personal circles.

With the foundation of the Royal Society of London and the Royal Academy of Science in Paris, the need for some better and wider means of communication became obvious and urgent. Conditions in England differed, however, from those in France. In England interest in scientific matters, in the new " philosophy," was actively centred in one group, which during the struggle between King and Parliament, met first in London, and then partly in London and partly in Oxford in separate assemblies, until their reunion in London after the Restoration to form " The Royal Society of London for the Improving of Natural Knowledge." But the *virtuosi* formed one group only [(2)]. In France, the very opposite conditions prevailed. In many French cities other than Paris there were independent groups of men interested in natural philosophy and meeting and joining in *conférences* and *sociétés* [(3)]. It was the fashion of the hour.

The foundation in 1666 of the Royal Academy of Science in Paris increased this interest in matters scientific, but there had already appeared in France in 1665 the first scientific journal, the 'Journal des Sçavans,' ancestor of all its kind [(4)]. Its editor, Denis de Sallo [(5)]

(1626–69), was a friend of Colbert, Louis XIV's Minister and prime mover in the establishment of the Académie Royale des Sciences in Paris. Descended from an ancient and noble family of Poitou and educated at the Collège des Grassins, he became a lawyer. He was an omnivorous reader of all kinds of books and he kept two secretaries to compile notes and transcribe extracts from the books that he read. Overwork brought ill-health and enforced inactivity; and de Sallo then thought that he might incorporate such material as he was led to abstract into a weekly journal of literary and scientific news, supposing that what had interested him would likewise interest others, especially the ever widening circle of amateurs of the "new philosophy."

The first number of the 'Journal' was published in Paris on Monday, January 5, 1665. The reader was advised that the journal would contain details of new books, obituaries, news of experiments in physics and chemistry, discoveries in the arts and in science, such as machines and the useful and curious inventions afforded by mathematics, astronomical and anatomical observations, legal and ecclesiastical judgments from all countries, and "enfin, on taschera de faire en sorte qu'il ne passe rien dans l'Europe digne de la curiosité des Gens de lettres. qu'on ne puisse apprendre par ce Journal." The founders, so the reader was informed, had given much consideration to the problem of whether the journal should appear annually, monthly or weekly. They had concluded in favour of weekly publication in view of the fact that news ages quickly: "Mais enfin ils ont creu qu'il devoit paroistre chaque semaine; parce que les choses vieilliroient trop, si on differoit d'en parler pendant l'espace d'un an ou d'un mois." They had been influenced also by the complaints of certain "persons of quality" that frequent issues of a journal were agreeable, whereas a whole volume of such items, from which the passage of time would have effaced the charm of novelty, might prove fatiguing [6].

De Sallo appeared as editor of the 'Journal' under the pseudonym of the Sieur de Hédouville, but it is not certain whether he sought anonymity or whether the name was taken from an estate he is said to have owned in Normandy. He had help in his editorial labours from Chapelain and probably from the Abbé Gallois, and possibly from others; and he allowed them complete freedom of opinion. However, this freedom led to criticism and raised an opposition that brought trouble upon the venture. After the appearance of the thirteenth number on March 30, 1665, de Sallo's privilege was withdrawn and the 'Journal' was suppressed, this step being said to be due, however, not to the complaints of authors who considered themselves aggrieved, but to Jesuit irritation at disrespectful Gallican comment on a decree of the Inquisition. De Sallo declined to continue publication with a censor. However, his resolute stand for independence had impressed Colbert, who remained friendly towards him and continued to consult him on the many subjects on which he was well informed. Towards the end of his brief life of forty-three years, de Sallo was given a financial post by Colbert because his private means had become seriously depleted through his generosity, his love of gaming and his haphazard life. Such was our first scientific editor.

After its suppression in 1665, the 'Journal' resumed publication on January 4, 1666, under the editorship of the Abbé Gallois, and was continued in 1685 by the Abbé de la Roque, in 1687 by Cousin and from 1702 to July 1792 by a committee. It was suppressed during the Revolution and revived in 1816. From 1665 to 1792, a series of one hundred and eleven volumes were published. The 'Journal' was reprinted in Amsterdam and also in Paris. It was imitated in other countries and even in France.

Although we are considering here the rise of the scientific periodical, and necessarily excluding the serial publications of the scientific academies, some reference must be made to the 'Philosophical Transactions' of the Royal Society of London. The Society had seen a copy of the 'Journal des Sçavans' and had discussed its contents and the problem of publishing a similar work; and they had concluded that something more "philosophical" was needed, excluding legal and theological items, but including more particularly accounts of the experiments made before the Society [7]. The first number of the 'Transactions' appeared in London on March 6, 1665. It was edited by Henry Oldenburg, one of the two Secretaries of the Society. It was almost wholly scientific in content; it did not cater for the interests of a widespread public of amateurs; it was a monthly and not a weekly. It was the official organ of the Society and thus the first of its kind. Its appearance marked a new development; for it was a medium for the publication of new observations and original work in science, mostly carried out by the Fellows of the Society, and it became the model on which all other published proceedings of the scientific academies have been fashioned. It reviewed books and gave space for the publication of differing scientific opinions by those engaged in similar experiments and studies. It was less amateur and more professional, if the latter term may be applied to the productions of an age when the professional scientist had not yet appeared on the scene.

However, although the 'Philosophical Transactions' was the Royal Society's official organ, it was financially the private venture of Henry Oldenburg and it cannot be said to have yielded him much in the way of mere profit. Its continuance was no more secure than that of the 'Journal des Sçavans,' although for different reasons. It is also to be remembered that some stimulus towards its production is definitely to be ascribed to the appearance of the 'Journal des Sçavans,' since a discussion of the 'Journal' by the Society led to the decision to publish the 'Philosophical Transactions'. The part played by Frenchmen in the foundation and development of the scientific periodical is inadequately recognized, as we shall see later on.

A word might be said in passing about Henry Oldenburg. He was born about 1615 in Bremen and died in Charlton, Kent, in 1677. He came to England about 1640 and, it is thought, remained here until 1648. In 1653 he came to England again to represent his native city on a matter of dispute with Cromwell and did not return to the Continent until 1657, when he acted as tutor and guardian to the Hon. Robert Boyle's nephew, Robert Jones, during a continental tour lasting until 1660. Returning to England, Oldenburg was appointed one of the two Secretaries of the Royal Society, the other being John Wilkins. Oldenburg carried on an extensive foreign correspondence on behalf of the Royal Society. Suspicion fell upon him during the Dutch War of 1665–67 and he was imprisoned in the Tower of London for two months. He was devoted to the new learning and enthusiastic in his labours to further the work of the Royal Society by " a commerce in all parts of the world with the most philosophical and curious persons to be found everywhere (8)."

The ' Journal ' and the ' Transactions ' gave two distinct models to scientific literature ; the former long influenced the development of the scientific periodical, until the rise of the journal specially devoted to one science only, while the latter became the pattern for the publications of the scientific academies that arose in greater numbers throughout Europe during the eighteenth century.

The ' Journal des Sçavans ' was imitated in Italy by the ' Giornale de Letterati ' (Rome), which continued to appear with a change of title from 1668 to 1697. The ' Philosophical Transactions ' seem to have been the model for the famous ' Acta Eruditorum,' the learned scientific monthly which first appeared at Leipzig in 1682. Leipzig was the centre of the German trade in books and one of the most important functions of the ' Acta ' was the announcement of new works of scholarship. It contained articles on the work of Europe's leading scientists ; it included in its field medicine and mathematics, law and theology ; it was the first German scientific periodical. Fifty volumes were published from 1682 to 1731, with ten supplementary volumes, followed by the ' Nova Acta Eruditorum ' in forty-five volumes from 1732 to 1782 with eight supplementary volumes. The first editor was Otto Mencke (1644–1707), and he was succeeded by his son and then by his grandson until 1754, when the editorship passed out of the family (9). The ' Acta ' will be remembered for many papers by Leibnitz, especially those dealing with his work on the calculus, which led to the controversy with Newton.

Mention might also be made of the ' Nouvelles de la république des lettres,' modelled on the ' Journal des Sçavans ' and edited by Pierre Bayle (1647–1706) during its first three years (Amsterdam, 1684–87). It was Holland's contribution to the cause of popular scientific periodical literature and continued until 1718.

The turn of the century in which the two types of scientific periodical had been established witnessed an important consequence of the reorganization of the Paris Academy of Science, namely, the publication of the first volume of the long series of 'Histoire et Mémoires' of the Academy, with many supplementary volumes, a series modelled on the 'Philosophical Transactions.'

The eighteenth century brought great developments in the scientific periodical. Bolton [10] lists seventy-four new journals originating between 1725 and the close of the century, excluding purely medical journals but including all other periodicals containing scientific material. However, it is clear from his bibliography, that by far the greater number of these originated after the middle of the century. Indeed, only five of them appeared before 1750, namely, 'Raccolta, d'opuscoli scientifici e filologici' (Venezia, 1728–57, 52 vols., continued as 'Nuova raccolta,' etc., Ferrara, 1755–87, 42 vols.). 'Le Pour et Contre' (Paris, 1733–40, 20 vols.), 'Bibliothèque Britannique, ou Histoire des ouvrages des savans de la Grande Bretagne' (La Haye, 1733–47, 25 vols., followed by the 'Journal Britannique,' La Haye, 1750–57, 24 vols.). 'Göttingische Zeitung von Gelehrten Sachen' (Göttingen, 1739–52, 14 vols., continued from 1753 to 1801 in 117 volumes under another title in association with the Göttingen Academy), and the 'Hamburgisches Magazin' (Hamburg, 1747–67, 26 vols., continued as the 'Neues Hamburgisches Magazin,' Hamburg and Leipzig, 1767–81, 20 vols.).

These new periodicals doubtless contributed to the development of the scientific journal, but their number, namely five, in a half-century, or, if we take the actual first years of publication, five between 1728 and 1750, does not indicate any striking change as compared with the developments in this kind of literature during the pioneering second half of the seventeenth century. However, the pace rapidly quickened. Of the remaining sixty-nine periodicals to appear between 1750 and 1798, nine appeared between 1750 and 1759, six between 1760 and 1769, and nine between 1770 and 1779; but from 1780 to 1789, there were twenty new scientific periodicals and from 1790 to 1798 a further twenty-five. It was therefore only during the last two decades of the eighteenth century that the number of journals was greatly increased; for forty-five of the sixty-nine new publications originated between 1780 and 1798.

Of the nine new periodicals originating between 1750 and 1759, probably the most important was the 'Commentarii de rebus in scientia naturali et medicina gestis,' a most useful and successful journal giving news of the work of the academies and details of new books. It was published in Leipzig, the first number appearing in 1752, and continued until 1798 in a long series of forty-one volumes, including supplementaries. With the 'Acta Eruditorum,' already mentioned, the 'Commentarii' are a mine of valuable information for the historian of science. In passing, it might be noted that of the nine new periodicals published in this decade six were German, two French and one Dutch. The predominance of the German contribution is again evident during the period from 1760 to 1769, when of the six new periodicals that appeared, five were German, the other being Italian. None of these six calls for any special comment here.

From 1770 to 1779, nine further new periodicals were published, seven of which were German, one French and one Italian. This decade witnessed a most important and significant development in the first appearance of the specialized scientific journal both in physics and in chemistry, namely, the 'Observations sur la Physique' (11), edited by the Abbé Rozier, and the 'Chemisches Journal' of Lorenz Crell. The 'Observations,' however, although largely concerned with physics, dealt extensively with chemistry as well and included sections on natural history and the arts. But the contents tended to become more physical, although in the early years of the journal it was the medium for the publication of many of Lavoisier's classical papers in chemistry in their original unrevised form before they were re-presented and re-read to the Paris Academy of Science (12). In the history of chemistry the journal is a most important source. In the development of the scientific periodical, its appearance is a landmark and the circumstances merit some consideration.

The full title of the journal was 'Observations sur la Physique, sur l'histoire naturelle, et sur les arts'; it was a monthly and the first number appeared in July, 1771. Eighteen numbers were published from that date to the end of 1772 in 12°. From January 1773, the format became 4° and continued so until the final issue of 1793. The first eighteen numbers are very rare and it had already become necessary to reprint them in 1777 in two 4° volumes under the title of 'Introduction aux Observations sur la physique,' etc. The volume for 1773, comprising the issues from January to June of that year, is numbered volume one; and in all forty-three volumes, usually two every year, appeared up to 1793.

The "Avis" to the reader prefacing volume one (1773) (13) explained that the publications of the academies were written in the languages of the different nations and printed several years after the memoirs included in them had been read before the academies; that during that interval facts that might prove of the greatest usefulness remained unknown; and, moreover, that the publications of the Academies, having become very numerous and therefore very costly, were often beyond the means of those to whom they might be advantageous. The writer went on to draw attention to other causes of delay: "Il semble qu'à mesure que le nombre des Savans s'est accru, la Correspondance, entre ceux des Nations différentes, a été rallentie. Chacun a cru sans doute que les Académies nationales seroient suffisantes, & qu'on en tireroit tous les secours nécessaires. La Constitution de ces Compagnies, formées par les Souverains, ayant admis des Correspondans étrangers, sembloit prévenir cette illusion, & remédier à l'inconvénient qui en est la suite; mais cette précaution, si sagement prise, n'a pas été justifiée par le succès.

"Il résulte de ce peu de communication, que les progrès des Sciences sont très-lents, que des Savans de deux Nations différentes, travaillent long-tems sur la même matiere, & qu'ils perdent un tems précieux pour acquérir une gloire qui devient à la fin problématique. Cet inconvénient est moindre que celui de travailler sur une matiere éclaircie par des travaux déja publiés, dont on n'a aucune connoissance. L'Auteur perd un tems qu'il auroit mieux employé pour le bien & la gloire de sa Patrie, si, en entrant dans la carriere qu'il vouloit parcourir, il eût eu sous les yeux le tableau actuel des connoissances physiques, & le terme où elles sont restées."

For these reasons it appeared necessary to publish a new journal: "Ces motifs ont fait desirer qu'un Ouvrage périodique, d'un débit sûr & animé, annonçât les découvertes que se font chaque jour dans les différentes parties des Sciences, soit par des Notices abrégées, soit par des Mémoires très-étendus, qui continssent le développement de toutes les preuves de ces découvertes, en traçant même la marche de l'esprit inventeur. On a pensé que ce moyen, le plus prompt pour la publication des découvertes nouvelles, accéléreroit également le progrès des Sciences, qui ne sont autre chose que la somme de ces découvertes.

"Telles ont été les raisons qui nous ont engagés à entreprendre ce Recueil; & nous les présentons avec d'autant plus de confiance, aux savans Etrangers, que ce sera leur ouvrage. Il est écrit dans une Langue, aujourd'hui celle de tous ceux qui ont reçu quelqu'éducation en Europe. L'Académie Royale des Sciences de Paris sentoit depuis long-tems l'importance de ce Recueil. Plusieurs de ses Membres le proposerent l'année derniere à peu-près suivant le plan auquel nous nous sommes attachés. Des raisons particulieres en ont empêché l'exécution totale."

But the new journal was not for "leisured amateurs" seeking entertaining reading under the illusion that they were initiated into scientific pursuits of which, in fact, they had no real knowledge: "Malgré l'accueil favorable que cet Ouvrage a reçu; malgré les éloges que les Savans lui ont donné, nous nous croyons obligés de circonscrire nos limites pour le rendre encore plus digne d'eux. Nous rejetterons en conséquence ci qui ne seroit que compilation indigeste, & dénuée de vues neuves & utiles. L'importance des matieres, la maniere dont elles seront présentées, nous décideront sur le choix des morceaux qui doivent être insérés dans ce Recueil. Nous n'offrirons pas aux Amateurs oisifs, des Ouvrages purement agréables, ni la douce illusion de se croire initiés dans les Sciences qu'ils ignorent." It was clear that the aims of the journal were serious and immediate and that popular science was not its object.

Moreover, the historical development of science was included in the plan: "Nous nous occuperons sur-tout de l'histoire des Sciences que nous embrassons dans notre plan; & c'est dans ces vues, que nous nous attacherons à rapporter les faits de la même espece, & les raisonnemens différens qu'ils auront fait naître. Cette maniere de voir & de comparer, présente un fond inépuisable d'instructions, que nous saisirons avec le plus grand soin; de sorte qu'on verra au premier coup d'œil, la suite des faits qui auront concouru à l'établissement d'une vérité importante [14]

" On ne sauroit trop inviter ceux qui veulent faire des progrès dans les Sciences, à rapprocher les connoissances transmises par les Savans de tous les siecles & de tous les pays. C'est un préalable nécessaire pour parvenir à de nouvelles découvertes, & ils doivent considérer, comme le premier pas qu'ils aient à faire, celui où les grandes Hommes, qui les ont précédés, ont terminé leurs travaux. La continuité des efforts des uns & des autres, forme cette union, & cet accord qui doit regner entre les Savans de tous les Pays pour étendre les limites des connoissances.

" C'est donc de cette réunion de travaux, de cette somme de connoissances que nous devons partir pour en acquérir de nouvelles, & pour donner à ce Recueil la consistance que son objet semble lui assurer. L'étendue de ce travail surpasseroit nos forces, si une société formée par des personnes uniquement occupées des Sciences utiles, ne daignoit concourir avec nous pour remplir le but que nous nous sommes proposé.

The journal was intended also to increase the circulation of scientific information of the first importance and the editor took Mersenne for his model : " Les vrais Savans n'ont pas la manie de faire des secrets de leurs découvertes ; amis de l'humanité, leur gloire est de lui être utile ; aussi, c'est à eux que nous offrons ce Recueil, comme un dépôt où ils ont droit de prendre acte de leurs découvertes. Nous les invitons à regarder notre Cabinet, comme celui du Pere Mersenne." And general and experimental physics, chemistry, medicine, agriculture and the arts and crafts were to be included in its province : " en un mot . . . . tout ce qui a rapport à l'observation & à l'expérience [(15)]." Memoirs written by authors in other languages would be translated into French or extracts would be published until further details would be forthcoming.

We have quoted at length from this historically important document because it reveals the state of affairs at this time with regard to the scientific periodical in general ; and its contents apply to the current situation and explain for a particular instance the circumstances that led to such an extraordinary increase in and development of scientific periodical literature during the last quarter of the eighteenth century.

The editor of the ' Observations ' deserves some mention. François Rozier was born at Lyon in 1734 and died there on the night of September 28–29, 1793. He was educated for the priesthood and revealed an unusual interest in scientific observation. Anxious to preserve his freedom, he resisted the attempts of the Jesuits to persuade him to join their Order. After his father's death in 1757, he managed an estate for his elder brother, agriculture being one of his constant interests and a subject on which he wrote extensively and became an authority as a result of the long experiments made on the estate that he superintended.

He became a friend of Rousseau and in 1771, now an Abbé, he acquired from Gautier d'Agoty an earlier journal called ' Observations sur l'histoire naturelle, sur la physique et sur la peinture,' which had first appeared in 1752 and had continued under changed and expanded titles not too successfully. Rozier was also a friend of the economist and physiocrat, Turgot, the Minister who in his short term of office brought about the reform of the French gunpowder-office under the organizing genius of Lavoisier. Rozier approved of the Revolution of 1789. He fell in the siege of Lyon, a devoted priest seeking by personal example to maintain the courage and assuage the sufferings of his people [16].

Until December of 1778, Rozier was sole editor of the ' Observations.' In 1779 he took as his assistant his nephew, J. A. Mongez, who was later sole editor from 1780 to 1785, when De la Métherie became editor. The ' Observations ' proved a striking success, because it achieved all its aims. Under the last editor the phlogiston theory was favoured and the papers on chemistry were not of a high standard, as the phlogiston theory was already being superseded by the new chemistry of Lavoisier except among some of the German chemists [17].

Much the same circumstances led Lorenz Crell to produce his various journals of chemistry ; and since, strictly speaking, the ' Observations sur la physique ' included subjects other than physics, Crell may be regarded as the founder of the first specialized scientific periodical. From 1778 to 1781 he published six volumes of his ' Chemisches Journal,' then thirteen volumes of ' Die neuesten Entdeckungen in der Chemie ' (1781–86), and two volumes of his ' Chemisches Archiv ' (1783), eight volumes of ' Neues Chemisches Archiv ' (1784–91) and one volume of ' Neuestes chemisches Archiv ' (1798), while his ' Chemische Annalen,' etc. ran to forty volumes from 1784 to 1803 with a ' Beitrage ' of six volumes from 1785 to 1799. A tireless correspondent and editor. In the meantime, however, the ' Annales de Chimie,' to be referred to presently, had been founded in 1789.

From 1780 to 1789 twenty new journals appeared. Some of these edited by Crell we have already mentioned. The others included eleven German and two French periodicals and the well-known English periodical, ' The Botanical Magazine, or Flower Garden Displayed ' (London, 1787, and in continuation), under the editorship of William Curtis (1746–99), who said of it that, whereas his ' Flora ' had brought him praise, the ' Magazine ' had brought him pudding [18]. This beautifully illustrated journal with its one hundred and fifty volumes and nearly ten thousand coloured plates continues to the present day and is a splendid example of our periodical literature. Its editors have included Jackson, Hooker, Prain, Stapf and A. W. Hill.

But the most important journal to appear between 1780 and 1789, was undoubtedly the ' Annales de Chimie,' edited by de Morveau, Lavoisier, Monge, Berthollet, de Fourcroy, de Dieterich, Hassenfratz and Adet, the representatives in France of the new chemistry of Lavoisier. Born in the year of the Revolution, 1789, the ' Annales ' outdistanced all its competitors. The ' Observations ' under De la Métherie favoured the old theory and in any case became more physical in content : Crell's periodicals survived for some years. But the ' Annales,' of which ninety-six volumes appeared in Paris between 1789 and 1815 together with three volumes of ' Tables des matières ' (1801, 1807, 1821), became the leading journal of chemistry. It continued under the title of ' Annales de chimie et de physique ' in seventy-five volumes from 1816 to 1840 and later resumed its original name to continue to our own day, oldest survivor among chemical periodicals and a lasting monument to the great editors who launched it in the shadow of revolution. Lavoisier, its most distinguished editor and its enthusiastic founder, saw only the first few years of its success as he fell a victim to the Terror in 1794.

The last decade of the century was remarkable for the publication of twenty-five new journals, thirteen German, three French and five English. Notable among these was Gren's ' Journal der Physik,' which first appeared in 1790 in Halle and Leipzig and which, strictly speaking, is the first journal specially devoted to physics, since Rozier's ' Observations ' included material from other scientific fields. The English journals included ' The Repertory of Arts and Manufactures,' first published in London in 1794, Nicholson's ' Journal of Natural Philosophy, Chemistry and the Arts,' which first appeared in 1797 in London and continued until 1813. The most notable journal to appear in this decade was, however, ' The Philosophical Magazine,' edited by Alexander Tilloch, the first number of which was published in London in 1798. In the physical sciences it was to become second in importance only to the ' Philosophical Transactions ' of the Royal Society of London. This month of June, 1948, marks the one hundred and fiftieth anniversary of its foundation. Its distinguished services to science throughout the world and its long and eventful history are described in another article in this number.

References.

(1) *Correspondance du P. Marin Mersenne*, ed. Tannery and de Waard, Paris, 2 vols. (1933–37).

(2) Even in the Cambridge of Newton's time it was found impossible to establish a "Philosophical Meeting," as Newton reported in a letter to Aston of Feb. 23, 1684/5, "for want of persons willing to try experiments" (C. R. Weld, 'History of the Royal Society,' etc., London, i. 305–6 (1848).

(3) See M. Ornstein, 'The Rôle of Scientific Societies in the Seventeenth Century, Chicago' (1928), and Harcourt Brown, ' Scientific Organizations in Seventeenth Century France (1620–1680)', Baltimore (1934). Attention is drawn to a recent valuable study by Miss R. H. Syfret, ' The Origins of the Royal Society ' (Notes and Records of the Royal Society of London, v. p. 75 (1948).

(4) Ornstein, *op. cit.*, pp. 198–209, and Brown, *op. cit.*, pp. 185–207.

(5) *Nouvelle Biographie Générale*, Paris, vol. 43, col. 189–91 (1864).

(6) Quoted from the Amsterdam reprint (1685) of the first volume of the *Journal* (1665). The writer has not been able to consult the original Paris issue of 1665.

(7) See a letter from Moray to Huygens, quoted by Harcourt Brown (*op. cit.*, p. 201).

(8) 'Dic. Nat. Biog.,' article on Oldenburg.
(9) *Nouvelle Biographie Générale*, Paris, vol. 34, col. 925–6 (1861).
(10) H. C. Bolton, 'A Catalogue of Scientific and Technical Periodicals (1665–1882),' etc., Washington, (1885), published in 'Smithsonian Miscellaneous Collections,' Washington (1887), vol. xxix. The number of journals listed is 5105. The second edition (Washington, 1897), carried the work to 1895 and included 8603 titles. The chronological tables given on pp. 1081ff of the second edition have been used in the present study. Bolton included in his bibliography all journals containing scientific material, but in the chronological tables he restricted the entries to purely scientific journals or seems to have done so, since he excluded the *Astronomisches Jahrbuch* (Berlin, 1774), the first successful astronomical journal and still surviving to our own time, presumably because the earlier volumes were ephemerides.
(11) The details of the title varied slightly from time to time, as may be seen by consulting Bolton (*op. cit.*, 2nd edn., p. 267, no. 2201).
(12) See Meldrum, 'The Eighteenth Century Revolution in Science—The First Phase,' Calcutta (1929), and McKie, 'Antoine Lavoisier,' London (1935).
(13) 'Observations,' etc., 1773, i. iii–vii.
(14) Many original memoirs from previous years were translated into French and given a greater currency in accordance with this policy.
(15) The influences of the *Encyclopédie* are evident in this passage.
(16) *Nouvelle Biographie Générale*, Paris, vol. 42, col. 827–30 (1863).
(17) See J. R. Partington and D. McKie, 'Historical Studies on the Phlogiston Theory,' Annals of Science, ii. 361 (1937) ; iii. 1 and 337 (1938), and iv, 113 (1939).
(18) See Samuel Curtis, 'William Curtis,' 1746–1799, etc., Winchester (1941), and an article by Hunkin in 'Endeavour,' v, 13 (1946).

SCIENTIFIC SOCIETIES TO THE END OF THE EIGHTEENTH CENTURY

By DOUGLAS McKIE, D.Sc., Ph.D.

THE first scientific societies [1] were founded in Italy in the sixteenth and seventeenth centuries. They were short-lived because of clerical opposition or suppression. The pioneer in their foundation was Giambattista della Porta (1538–1615), who in 1560 organized the Academia Secretorum Naturæ from a group of friends interested in experiments and meeting in his house in Naples. Membership was restricted to those who had a discovery to report or some advance in physics to communicate: and the members called themselves the "Otiosi." However, in a short time, Della Porta was accused of magic and sorcery, and summoned to Rome to defend himself; he was acquitted, but ordered to close the academy.

The second scientific society, the Accademia dei Lincei, was formed in Rome about 1600 or 1603 and reorganized in 1609. It was founded by the Prince Fredrigo Cesi and its members included Della Porta, Galileo, Pieresc, Fabio Colonna and others. After 1630, when its patron died, and especially after Galileo's condemnation in 1633, the continuation of the work of the academy became difficult, although it did not come to an end until 1657. This second academy had, however, succeeded where the first had failed; for a number of works were published under its auspices, notably Galileo's 'Il Saggiatore' (Rome, 1623) [2].

Italy, cradle of the Renaissance, saw also the foundation of the third scientific society, the famous Accademia del Cimento, or Academy of Experiment, at Florence in 1657, under the patronage of the Grand Duke Ferdinand II of Tuscany and his brother Leopold, of the Medici family, and friends and pupils of Galileo, Viviani and Torricelli. After ten years of brilliant activity, this third academy came to an end in 1657; in that same year, Leopold de Medici received the cardinal's hat that he had long coveted, and clerical opposition is said to have succeeded once again in terminating the work of those who sought to engage in the study of natural philosophy by means of experiments. The researches of the academy were published in Florence in 1666 under the title of 'Saggi di Naturali Esperienze, etc.', a second edition following in 1667. This first classic of modern experimental physics enshrines the work of the nine members of the academy [3]; it was translated into English and into Latin, and later into French, and has been reprinted several times [4].

With the extinction of the Accademia del Cimento, leadership in physics passed from Italy to England and to France; it was a time marked also by the transition from Galileo to Newton. Meanwhile, developments had occurred in Germany. The Societas Ereunetica was founded by Joachim Jungius (1587–1657) at Rostock in 1622 for the study of experimental science, but survived for no longer than two years. The Collegium Naturæ Curiosorum, founded by the physicians of Schweinfurt in 1652 at the instance of Lorenz Bausch (1605–65), burgomaster and town-physician, was mainly concerned with medicine as related to science: in 1672 it received official recognition by the Emperor Leopold and became the Academia Caesareo-Leopoldina naturæ curiosorum. The second purely scientific society in Germany was the Collegium Curiosum sive Experimentale, founded at Altdorf in 1672 by Johan Christopher Sturm (1635–1703), professor of mathematics in the University of Altdorf, on the model of the Accademia del Cimento and for the experimental study of physics. Its importance diminished with the rise of the Prussian Academy in Berlin and the Bavarian Academy in Munich in the course of the eighteenth century.

Seventeenth-century France had shown a widespread interest in the "new philosophy" in which experiment and observation took the place of text and authority. Many societies and groups were formed both in Paris and in other cities for the purpose of indulging an interest in scientific matters and problems. Much of this interest was of an amateur kind, but it was active and general; and the various groups and societies attracted many cultured and scholarly members. In the period from 1620 to 1680, so thoroughly and admirably studied by Harcourt Brown [(5)], we find accounts of the groups in Paris associated with de Thou and with the brothers Pierre and Jacques Dupuy, the meetings held by Descartes in his lodgings about 1626, the conferences of Renaudot held in the Bureau d'Adresse which began in 1633 and continued until 1624, the meetings in the cell of Marin Mersenne (1588–1648), the Minorite father, the assemblies from 1654 to 1664 in the house of Montmor, the conferences of Henri Justel (1620–93) beginning in 1664, and the academy of the Abbé Bourdelot (1610–85), while in the provinces there were many societies and academies.

In England a very different state of affairs prevailed. Instead of many groups, there was only one. Its meetings from about 1645 onwards led eventually to the formation of the Royal Society of London on November 28, 1660, and its incorporation by Royal Charter on July 15, 1662 [(6)]. The successful foundation of the Royal Society proved to be one of the factors in the formation of the Académie Royale des Sciences of Paris in 1666. Both societies, or, as they should be more properly called, national academies, prospered. The close of the seventeenth century found them securely established and actively pursuing their aims. Their history from their formation to the end of the seventeenth century is too well-known to detain us here.

The eighteenth century therefore opened with two active national academies concerned with science and with the publication of new facts and discoveries, the Royal Society of London and the Académie Royale des Sciences of Paris. The Italian societies had died away ; the German societies had not been entirely successful and Germany needed the enthusiasm and genius of a Leibnitz for the foundation of a national academy of science. However, the seventeenth century had given two great models to the eighteenth century and national academies soon increased and multiplied : and independent and specialist societies were to appear as the century progressed in many countries in Europe, while a beginning was made in America.

In 1700 the Paris Académie opened with a new constitution and was completely reorganized. Its membership was increased from twenty to seventy. Official and continuous publication of its work was now undertaken and the long series of 'Histoire et Mémoires' began to appear in 1702, when the volume for 1699 was published as the first of the series. The 'Philosophical Transactions' of the Royal Society of London and the 'Histoire et Mémoires' of the Paris Royal Academy of Science were the outstanding official organs of the national scientific academies of the eighteenth century. The Académie and all other academies of the old régime were suppressed during the Revolution in 1793, and the Institut National was founded to replace them in 1795. In 1816 the Académie des Sciences was revived as part of the Institut.

In Germany in this same year of 1700 a national academy of science was at last established. Its foundation was essentially the result, not of the interest of a group of natural philosophers, but of over thirty years of compaigning and perseverance on the part of Gottfried Wilhelm von Leibnitz (1646–1716). Modelled on the academies of London and Paris, the new society received its charter on July 11, 1700, as the Societas Regia Scientiarum. Its official organ, 'Miscellanea Berolinensia,' first appeared in 1710, twelve of the sixty memoirs in it being contributed by Leibnitz. However, Frederick I had been succeeded by Frederick William I, and Leibnitz was not popular with the new monarch, who regarded scientific work as mere trifling. Moreover, relations between Leibnitz and the Academy had deteriorated. The second volume of the 'Miscellanea' did not appear until 1723, seven years after his death. Eight volumes were published between 1710 and 1744.

Leibnitz had a hand in the formation of other societies. He tried to found an academy in Dresden, but a war intervened. After his experience in Berlin, he may have become aware of the difficulty of obtaining adequate financial support for such societies ; and his proposals for an academy in Saxony included the suggestion that money was to be provided from the tobacco monopoly, a source of revenue which has sometimes been tapped increasingly elsewhere for purposes that do not always arouse much enthusiasm. Leibnitz achieved a greater success in Russia. In 1711 he met Peter the Great, who was much impressed with his very ractical ideas for the improvement of national education and for the liffusion and increase of scientific knowledge. Although the Academy of St. Petersburg was not founded until twelve years after Leibnitz had died, it was derived directly from his conversations with Peter the Great. In Vienna, too, in 1710–12, Leibnitz, but for Jesuit opposition, would have succeeded in instituting an academy, the foundation of which was delayed until 1846.

The plan for the St. Petersburg Academy was approved by the Emperor early in 1724, but his sudden death arrested further progress. In December of 1725, the Empress Catherine I formally established the Academy and the first meeting was held on the 27th of the month. Throughout the eighteenth century, much valuable work was carried out by the academicians, especially on the natural resources of Russia. The history of these activities is still largely unknown. On the strictly scientific side, it need only be mentioned that many members of other nations were invited to join the Academy, which during this period included among its members Daniel Bernoulli, Krafft, Richmann, Euler and many others. Michail Lomonossov (1711–65), a versatile genius and forerunner of Lavoisier in rejecting the phlogiston theory in chemistry, was a member from 1741 until his death. The first publications of the Academy appeared in 1728 as 'Commentarii Academiæ Scientiarum Imperialis Petropolitanæ' and continued until 1747, when the title was changed to 'Novi Commentarii, etc.', which in turn became 'Acta Academiæ, etc.', in 1777. As in the Paris Académie des Sciences, the members, whose number was limited to fifteen, were salaried ; in addition, they received the title of professor in the various subjects in which they worked.

In Sweden, the Kongliga Svenska Vetenskaps Akademien arose from a society, the Collegium Curiosorum, which first met in 1739 and included Linnæus among its six members ; in 1741 it was formally established as the Royal Swedish Academy, and its work and publications became well-known during the eighteenth century, forty volumes of the first series having appeared up to 1779. In Denmark, the Kongelige Danske Videnskabernes Selskab, the Royal Academy of Science, was constituted by Christian VI at Copenhagen in 1743.

Before the middle of the century Benjamin Franklin had urged the formation of a scientific society in America in a pamphlet entitled 'A Proposal for Promoting Useful Knowledge among the British Plantations in America,' dated from Philadelphia on May 14, 1743 [7]. It contained detailed suggestions of a practical kind, as might be expected from its author, and among these it recommended "That One Society be formed of Virtuosi or ingenious Men residing in the several Colonies, to be called *The American Philosophical Society* who are to maintain a constant Correspondence. That *Philadelphia* being the City nearest the Centre of the Continent-Colonies, communicating with all of them northward and southward by Post, and with all the Islands by Sea, and having the Advantage of a good growing Library, be the Centre of the Society. That at *Philadelphia* there be always at least seven Members, *viz.*, a Physician, a Botanist, a Mathematician, a Chemist, a Mechanician, a Geographer, and a general Natural Philosopher, besides a President, Treasurer and Secretary." It concluded with the statement that "*Benjamin Franklin*, the Writer of this Proposal, offers himself to serve the Society as their Secretary, 'till they shall be provided with one more capable." The American Philosophical Society Held at Philadelphia for Promoting Useful Knowledge was formed shortly afterwards. It celebrated its bicentenary in 1943. Franklin was President from 1769 to 1790 and Thomas Jefferson from 1797 to 1815. The Society's. 'Transactions' began to appear in 1771; its 'Proceedings' date from 1838. Another scientific society was formed in America during this period, namely, the American Academy of Arts and Sciences, founded at Boston in 1780. Its 'Memoirs' date from 1785, its 'Proceedings' from 1846.

Shortly after the middle of the century a private scientific society was again formed in Italy at Turin in 1757; and 1783 it became the Reale Accademia della Scienze, or Royal Academy of Science. The famous Accademia dei Lincei was not revived until the nineteenth century, when the Reale Accademia dei Lincei was constituted in 1870. The Lisbon Academy, the Academia Real das Sciencas, originated in 1779 and was reorganized in the nineteenth century. In Spain an academy was formed in 1657 in Madrid modelled on Della Porta's Academy at Naples; it was reorganized in 1847 with a constitution corresponding to that of the Paris Academy.

Provincial academies were formed in France in considerable numbers during the eighteenth century, namely, in Montpellier, Bordeaux, Dijon and other cities. The earliest society in Brussels dates from 1769 and its interests were literary; scientific academies in the Low Countries date from the nineteenth century. In Germany academies were formed in Göttingen in 1750, Erfurt in 1754, Mannheim in 1755 and Munich in 1759.

Enough has been said in the space of this short paper to indicate the rise and development of the scientific academies in the eighteenth century. Space prevents a full account of the proliferation of provincial scientific societies throughout Europe, but we shall give some brief account of some of the interesting and important, and some of the less important but none the less interesting, scientific societies and specialized scientific societies that arose in Great Britain during the eighteenth century. The detailed history of this phase of the history of science in Europe awaits further study.

Attempts had been made in 1683 to form a society in Dublin after the pattern of the Royal Society of London. A year later the society was established with Sir William Petty as President and the members numbered thirty-three at the close of that year. Though in correspondence with the parent society in London, the Dublin Society did not prosper; Ireland was disturbed by the political changes that occurred shortly afterwards [8]. The Royal Irish Academy was formed late in the eighteenth century from a society in Dublin organized about 1782; its 'Transactions' first appeared in 1788.

In Scotland affairs went differently despite the political uncertainties of the eighteenth century. There was a philosophical society in Edinburgh in 1705 in communication with the Royal Society of London. Later the Earl of Morton supported the plans of Maclaurin, the mathematician, for extending the activities of a society formed in Edinburgh in 1731, and in 1739 the Philosophical Society of Edinburgh was accordingly instituted. Its publications appeared in three volumes as 'Essays and Observations, Physical and Literary, Read before a Society in Edinburgh, and Published by them' (Edinburgh, 1754–71). The second volume, published in 1756, contained the classical researches of Black on the mild and caustic alkalis, an extended study read to the Society in 1755 of work submitted for the degree of M.D. (Edinburgh) in 1754. In 1783 the Society became by charter the Royal Society of Edinburgh [9]. Its roll of Fellows contains many names famous in the history of science. Mention might also be made of the society or club formed by the professors in the University of Glasgow during the second half of the eighteenth century. It was to this society on April 23, 1762, that Joseph Black announced his classic discovery of the latent heat of fusion of ice. Black refers to this body as "the Philosophical Club, or Society of Professors and others in the University of Glasgow" [10].

As the century drew towards its last two decades, Richard Watson (1737–1816), absentee Bishop of Llandaff and Regius Professor of Divinity in the University of Cambridge, lately Professor of Chemistry in that university, wrote in his 'Chemical Essays' in 1781:—"The Philosophical Transactions at London, the Histoire de l'Academie Royale des Sciences at Paris, the Saggi d'esperienze di Acad. del Cimento at Florence, the Journal des Scavans in Holland, the Ephemerides Academiæ naturæ curiosorum in Germany, the Acts of the Academy of Copenhagen, and the Acta eruditorum at Leypsic; all these works began to be published within the space of twenty years from 1665, when our Royal Society first set the example, by publishing the Philosophical Transactions. To these may be added, the works of the academies of Berlin, Petersburgh, Stockholm, Upsal, Bononia, Bourdeaux, Montpelier, Gottingen, and of several others which have been established within the course of the present century. Near a thousand volumes have been published by these learned societies within less than 120 years" [11]. Societies and academies had multiplied greatly; but, as there was still a need for further periodicals in this period and for the specialized periodical, so also was there need for further societies and for specialized societies.

Recently Kendall [12] has drawn attention to some early chemical societies in Scotland. He showed that the Chemical Society mentioned in Joseph Black's papers was formed from students attending Black's lectures on chemistry in Edinburgh in 1785; that there was a Chemical Society in the University of Glasgow as early as 1786; and that in the early months of 1800 another chemical society was formed in Edinburgh. Wilson [13] had already drawn attention to the formation of a second chemical society in Glasgow in 1798. Until Kendall brought these facts to light, it was generally considered that the oldest chemical society was the Chemical Society of Philadelphia, which was founded by James Woodhouse in 1792.

A Botanical Society was founded in London in 1721 by Johann Jakob Dillen or Dillenius (1681–1747), Sherardian Professor of Botany in the University of Oxford, and John Martyn (1699–1768), son of a London merchant and amateur botanist and entomologist. The Society met at the Rainbow Coffee House in Watling Street, and later in a private house, at six o'clock on Saturday evenings, but it lasted for only five years. In 1745 the Aurelian Society for the study of insects began to meet in the Swan Tavern in 'Change Alley, but the fire in Cornhill in 1748 destroyed its library and its collection, and the Society appears to have perished also in the disaster. A second Aurelian Society lasted from 1762 to 1766. From 1780 to 1782 there was a Society of Entomologists of London. In 1782 on October 21, " at Mr. Dean's, the Corner House by the Turnpike, Pimlico," the Society for Promoting Natural History was founded. Later the members met in the Black Bear in Paccadilly, in " Greenwoods " or Leicester House in Leicester Square, in the house of one of its members, George Spence, in Pall Mall, and in its own rooms at 19 Warwick Square. Later it met in the York Coffee House in St. James's Street. Meetings were held at seven in the evening every month " on the Monday next following the Full Moon," this precaution being taken on account of the darkness and insecurity of the streets of eighteenth-century London. The historian of the Linnean Society records that the charges at the York Coffee House were invariably 10*s*. 6*d*. for the rooms and £1. 1*s*. for sundries, and adds that " it is not mentioned what the sundries were, but it is improbable that they consisted purely of ink." The general dissatisfaction of the abler members of the Society with its methods and its work led to the formation of the Linnean Society in 1788 by James Edward Smith (1759–1828), Samuel Goodenough (1743–1827) and Thomas Marsham (1747 or 8 –1819). Smith will be remembered for acquiring in 1784 the Linnean Herbarium, library and manuscripts, which subsequently became the property of the Society and remain its proudest possession. The foundation of the Linnean Society in 1788 marks the successful establishment of the first specialist scientific society in England. The Society for Promoting Natural History generously handed its funds, books and collections to the Linnean Society some years later. The history and reputation of the Linnean Society since its foundation are too well-known to need any further reference here [14].

The most important provincial scientific society in England was formed in Manchester in the later part of the eighteenth century. For several years before its foundation, a weekly club had been organized for those with an interest in literature and philosophy, *i. e.*, natural philosophy ; the club became in 1785 the Literary and Philosophical Society of

Manchester. The first volume of the 'Memoirs' of the Society was published in Warrington in 1785. The Preface to this volume brings vividly before us the thought and feeling of another age [15]: "The numerous Societies, for the promotion of Literature and Philosophy, which have been formed in different parts of Europe, in the course of the last and present centuries, have been not only the means of diffusing knowledge more extensively, but have contributed to produce a greater number of important discoveries, than have been effected in any other equal space of time.

"The progress that has been made in Physics and the Belles Lettres, owes its rapidity, if not its origin, to the encouragement which these Societies have given to such pursuits, and to the emulation which has been excited between different academical bodies, as well as among the individual Members of each institution. The collecting and publishing the more important communications which have been delivered to them, have saved from oblivion many very valuable discoveries, or improvements in arts, and much useful information in the various branches of science. These their modest authors might have been tempted to suppress, but for the respectable sanction of societies of men of the first eminence and learning in their respective countries, and the easy mode of publishing, which their volumes of Transactions afford."

But Manchester, then as now, had no love for what might be called metropolitan parochialism: "Though, in France, Societies for these purposes have been instituted in several of the provinces, in England, they have almost been confined to the Capital; and however great have been the advantages resulting from the researches of the learned bodies, who are incorporated in London, it seems probable, that the great end of their institutions, the promotion of arts and sciences, may be more widely extended by the forming of Societies, with similar views, in the principal towns in this kingdom.

"Men, however great their learning, often become indolent, and unambitious to improve in knowledge, for want of associating with others of similar talents and acquirements: Having few opportunities of communicating their ideas, they are not very solicitous to collect or arrange those they have acquired, and are still less anxious about the further cultivation of their minds.—But science, like fire, is put in motion by collision.—Where a number of such men have frequent opportunities of meeting and conversing together, thought begets thought, and every hint is turned to advantage. A spirit of inquiry glows in every breast. Every new discovery relative to the natural, intellectual or moral world, leads to a farther investigation; and each man is zealous to distinguish himself in the interesting pursuit.

"Such have been the considerations that have led to the institution of the Literary and Philosophical Society of Manchester."

We have quoted this at length, because it is the vindication of all such societies in the period under consideration and above all in the expanding years with which the century closed. This great provincial society has had a long and distinguished history; its roll of members has borne many famous names and we need recall only one, that of John Dalton, who read to the society his first papers on the atomic theory and exhibited at its meetings the first table of atomic weights. The whole world of science regrets that war brought complete destruction to the buildings and historic treasures of the oldest learned society in England outside London.

Birmingham, too, had its society, the Lunar Society, whose members were sometimes referred to as the "Lunatics" [16]. The name was derived from their practice of meeting when the moon was full, in order to have light for the journey home, as we have seen was the custom in London in the Society for Promoting Natural History. The Lunar Society existed as early as 1768 and was founded by Matthew Boulton. It included among its members Erasmus Darwin, Samuel Galton, Richard Lovell Edgeworth, Thomas Day (author of 'Sandford and Merton'), James Keir, William Withering, James Watt and Joseph Priestley; it met at two in the afternoon for dinner and the meetings lasted until eight; many distinguished guests were entertained, and among them were Wedgwood, Banks, William Herschel, Smeaton, Samuel Parr, Afzelius, Solander, De Luc and others. The influence of the society in the stimulation of scientific inquiry was outstanding in that remarkable age; and it persisted long after the meetings had to be abandoned on account of there being too few survivors to continue them.

A different kind of society was the Spitalfields Mathematical Society [17], founded in 1717 by John Middleton and originally said to be formed of Spitalfields weavers, later of tradesmen and professional men of the neighbourhood. Little is known of the early history of this group, but it seems that every member, on penalty of being fined twopence, had to answer any mathematical or philosophical question that was put to him. The members were provided with pipes and porter to improve the occasion and doubtless to help them to concentrate on their problems. Philosophical lectures were given and charges for admission to them were made. In 1799 this activity brought the society almost to extinction; for a gang of informers laid an information against the members that they had taken money for an unlicensed entertainment, namely, a philosophical lecture. A voluntary guarantee fund of £254 was raised among the forty or fifty members in a few days and one of them took over the defence without fee; but scandal had been brought upon them and their lectures were not so well attended in future, while the informers were legally free to repeat the charge when they wished. After this, the supply of pipes and porter had to be left to the members; and it was found that "the small number attending on Saturday evenings arises in great measure from their not drinking in common," whereupon the custom was resumed. Every member had to lecture in turn and medals were awarded. Much apparatus was collected and also a library of over 2500 volumes, including many rare books. There were no distinguished names among its members. In 1846, after a life of one hundred and thirty years, the society was absorbed by the Royal Astronomical Society, who received the library and other items, including several valuable portraits.

In London in 1793 Bryan Higgins, encouraged by a number of friends with " a taste for Philosophical Experiments and Conversations," founded a new society for such experiments and conversations. The first meeting was held in Higgins's House on January 25, 1794, at 8 o'clock, and the Society for Philosophical Experiments and Conversations was formally instituted. Weekly meetings were held at the same place " during the Session of Parliament." Higgins lent his apparatus for the experiments. The annual subscription was five guineas. Fifty intending members had subscribed before the first meeting. Others were elected later. The Chairman was Field-Marshal Conway ; the " Didactic Experimenter " was " Dr. Bry. Higgins " ; and " Mr. Thomas Young " was one of three " Assistant Experimenters." The procedure at the meetings consisted in the performance of experiments followed by discussion of their results. Many of Lavoisier's classic experiments were repeated ; and the members appear to have been mainly interested in the new chemistry of Lavoisier, especially the problems of combustion and respiration, of the composition of the air and the properties and measurement of caloric. Twenty-one meetings were held, the last on June 21, 1794. A committee of publication met on July 3 of that year ; and at this point the activities seem to have ended [(18)].

Towards the close of the century two societies were formed in London, the Askesian Society in March 1796 and the British Mineralogical Society in April 1799. The latter combined with the former in 1806, and some of the members of the combined society instituted the Geological Society in 1807 [(19)]. In the nineteenth century many more specialist societies arose throughout the world ; but it was the eighteenth century that gave the vital impetus to a movement that is characteristic of our time.

In closing, we can make only brief reference to two other developments at the very close of the century. Between 1794 and 1798 Beddoes had succeeded in establishing with the aid of subscriptions, mostly from Wedgwood and Watt, the Pneumatic Institution at Clifton for the experimental study of the medicinal effects of gases. The institute for research had unobtrusively made its first appearance. The name and genius of Humphry Davy is associated with this beginning and with the other innovation that followed shortly in the foundation in London in 1799 by Sir Benjamin Thompson, Count Rumford, of another institution, chartered in 1800 as the Royal Institution for the purpose of diffusing scientific knowledge and its applications to everyday life. Its subsequent history has been one of research and classic discoveries in chemistry and physics.

References.

(1) See M. Ornstein, 'The Rôle of Scientific Societies in the Seventeenth Century,' Chicago, 1928, and 'Encyclopædia Britannica,' 11th ed., article " Academies " (section I.).

(2) Other works published were Galileo's book on sun-spots, 'Istoria e Dimostrazioni intorno alle Macchie Solari, etc.', Rome, 1613, and Oviedo's 'Thesaurus Mexicanus,' Rome, 1651.

(3) The nine members were Borelli, Candido and Paolo del Buono, Magalotti, Marsili, Oliva, Redi, Renaldini and Viviani.

(4) An English translation of the "Saggi" by Richard Waller was published in London in 1684 as 'Essayes of Natural Experiments made in the Academie del Cimento, etc.' A Latin version by van Musschenbroek, 'Tentamina experimentorum, etc.' was published in Leyden in 1731; and a French translation appeared in 1755. The work was reprinted in Italian in the 'Atti e Memorie inedite, etc.' in Florence in 1780 and again in 1841.

(5) Harcourt Brown, 'Scientific Organizations in Seventeenth Century France (1620–1680),' Baltimore, 1934.

(6) We draw attention to a recent valuable study on "The Origins of the Royal Society" by Miss R. H. Syfret in 'Notes and Records of the Royal Society of London' (v. p. 75, 1948).

(7) Quoted from the reproduction of this very rare pamphlet in Proc. Amer. Phil. Soc. lxxxvii. pls. iii. and iv. (1943). This number contains an interesting paper by van Doren (pp. 277 ff) on the beginnings of the American Philosophical Society, with a reproduction of the beautiful portrait of Franklin by Peale (pl. ii.).

(8) C .R. Weld, 'History of the Royal Society, etc.', London, i. pp. 299–302, 383–84 (1848).

(9) *Ibid.* i. p. 383, and ii. p. 24. See also Kendall, 'Endeavour,' v. p. 54 (1946).

(10) McKie and Heathcote, 'The Discovery of Specific and Latent Heats,' London, pp. 19 and 35 (1935).

(11) "Chemical Essays', 1st edn., London, i. pp. 35–36 (1781). Watson refers to the 'Journal des Scavans' as appearing in Holland, and presumably he was aware only of the Amsterdam reprint, which would explain why he states that the Royal Society "first set the example" by publishing the 'Philosophical Transactions,' whereas the 'Journal' preceded the 'Transactions' by two months (January to March, 1665).

(12) "Some Eighteenth-Century Chemical Societies" ('Endeavour,' i. p. 106, 1942). This paper mentions also a chemical society in London in 1782 meeting once a fortnight in the Chapter Coffee-house. See also Sir Wm. Ramsay's 'Life and Letters of Joseph Black, M.D.', London, pp. 110–11 (1918); and Kendall, J. Chem. Education, xii. p. 565 (1935).

(13) 'Annals of Science,' ii. p. 451 (1937).

(14) For the details of this paragraph the writer is indebted to A. T. Gage's 'History of the Linnean Society of London,' London, pp. 1–10 (1938).

(15) Mem. Lit. & Phil. Soc. Manch. i. pp. v–vii (1785).

(16) See T. E. Thorpe, 'Joseph Priestley,' London, 94 ff (1906), and Anne Holt, 'A Life of Joseph Priestley,' London, 127 ff. (1931).

(17) For information about the Spitalfields Society the writer is indebted to R. A. Sampson's contribution to the 'History of the Royal Astronomical Society 1820–1920,' ed. J. L. E. Dreyer and H. H. Turner, published by the Royal Astronomical Society, London, pp. 99–104 (1923).

(18) 'Minutes of the Society for Philosophical Experiments and Conversations,' London (1795). The publication committee of the Society resolved to publish this account of the work done during 1794. No further volumes appeared.

(19) See H. B. Woodward, 'The History of the Geological Society of London,' London, pp. 6–10 (1907).

THE TEACHING OF THE PHYSICAL SCIENCES AT THE END OF THE EIGHTEENTH CENTURY

By F. SHERWOOD TAYLOR, Ph.D.

An examination of scientific education at any period resolves itself into two enquiries, concerning the manner in which scientists of that period were in fact educated, and concerning the means of education that were available. Scientists today pass, with rare exceptions, through the officially provided channels of secondary school, or its equivalent, and university, but in the eighteenth century a considerable proportion of the great men of science received no formal instruction in the subject they made their own, or attended a course in it as a discipline ancillary to some other subject, normally medicine.

(1) *The manner of Education of the Scientists born between* 1760 *and* 1800.

For the purpose of this paper, the author selected the names of 68 persons who were born between 1760 and 1800, and ascertained the steps by which they took up the study of the subject in which they became famous. The results may be expressed in the form of the tables, given below. It is not claimed that the classification of these scientists is unimpeachable, for in some cases there is doubt concerning the category to which they may be most properly assigned.

Table I.—Men of science who received no formal higher education.

Name	Year of Birth	Remarks
Ampère, A. M. ..	1775	Educated himself privately in mathematics.
Baily, F.	1774	Became an astronomer in middle life.
Bessel, F. W. ..	1784	Went as super-cargo to study mathematics and navigation.
Brunel, M. I. ..	1769	Educated for priesthood.
Chevreul, M. E. ..	1786	Enters Vauquelin's manufactory of chemical products.
Daguerre, L. J. M. ..	1789	Painter.
Dalton, J.	1766	Self-educated.
Daniell, J. F. ..	1790	Business career ; sugar refining.
Faraday, M.	1791	Attended some private lectures.
Loudon, J. C. ..	1783	Apprenticed to a nursery-man.
Niépce, J. N. ..	1765	Sub-lieutenant in army.
Smith, W.	1769	Became a surveyor, learnt geology by observation.
Stevenson, G. ..	1787	No education.
Sturgeon, W. ..	1783	No education ; became a private soldier.
Wheatstone, C. ..	1802	Musical instrument maker.

TABLE II.—Men who received a higher education in non-scientific subjects.

Name	Year of Birth	Remarks
Amici, G. B.	1786	Engineer-architect at Bologna.
Avogadro, A. ..	1776	Jurisprudence.
Brewster, D.	1781	Studied for ministry and incidentally heard scientific lectures at Edinburgh.
Buckland, W. ..	1784	Theological studies.
Leslie, J.	1766	Theological studies.
Schweigger, J. S. C. ..	1779	Philosophical and classical studies.

TABLE III.—Men who received a predominantly mathematical education, including those whose training was military, for artillery or engineers.

Name	Year of Birth	Remarks
Arago, D. F. J. ..	1786	École polytechnique.
Biot, J. B.	1774	Artillery.
Bolyai, W. (pére) ..	1775	Mathematical education at Göttingen.
Fourier, J. B. J. ..	1768	Military School.
Fresnel, A. J. ..	1788	École polytechnique: became an engineer.
Gauss, K. F.	1777	Mathematical education at Göttingen.
Herschel, J. F. W. ..	1792	Cambridge, Senior Wrangler.
Malus, E. L.	1775	Military engineering: École polytechnique.
Nobili, L.	1784	Artillery Officer.
Ohm, G. S.	1787	Mathematics: University of Erlangen.
Quetelet, L. A. J. ..	1796	Mathematical education.
Sedgwick, A.	1785	Cambridge Wrangler: knew little or no geology until made Professor.
Talbot, W. H. F. ..	1800	Cambridge, 12th Wrangler.
Whewell, W.	1794	Cambridge, 2nd Wrangler.
Woodhouse, R. ..	1773	Cambridge, Senior Wrangler.

TABLE IV.—Men who attended general scientific courses, other than medical or mathematical, at a university.

Name	Year of Birth	Remarks
Buch, L. von ..	1774	Studied geology under Werner at Freiberg.
Chamisso, L. C. A. ..	1781	Royal College of France.
Cuvier, G. L. C. F. D.	1769	Course in philosophy, mathematics and natural science at Stuttgart.
Jameson, R	1773	Studied geology under Werner at Freiberg

TABLE V.—Men who began their scientific career through medicine or pharmacy.

Name	Year of Birth	Remarks
Beddoes, T.	1760	Medical education. London and Edinburgh.
Berzelius, J. J. ..	1779	Medical and chemical studies at Upsala.
Brande, W. T. ..	1788	Apprenticed to an apothecary with a view to medicine.
Brongniart, A. ..	1770	École de Medecin, but previously École des Mines.
Brown, R.	1789	Marischal College, Aberdeen, then medical studies at Edinburgh.
Candolle, A. P. de ..	1778	Medical studies, unwillingly undertaken.
Carlisle, A.	1768	Medical pupil : surgeon. Electrical researches incidental.
Cooper, A.	1768	Apprenticed to surgeon : also medical school at Edinburgh.
Daubeny, C. J. B. ..	1795	Medical studies at Edinburgh.
Davy, H.	1778	Apprenticed to a surgeon : then employed at Dr. Beddoes' Pneumatic Institution.
Dobereiner, J. F. ..	1780	Studied pharmacy, then set up chemical manufactory.
Dulong, P. L. ..	1785	Abandoned a medical career in favour of chemistry.

TABLE V.—(*cont.*)

Name	Year of Birth	Remarks
Dumas, J. B. A. ..	1800	Apprenticed to an apothecary: then moved to Geneva.
Dutrochet, R. J. H. ..	1776	Military marine: medical studies from 1802.
Gmelin, L.	1788	Medicine and chemistry; Göttingen, Tübingen, Vienna.
Henry, J.	1797	Studied for medicine.
Laënnec, R. T. H. ..	1781	Medical career.
Oersted, J. C. ..	1777	Pharmacist's apprentice.
Oken, L.	1779	Medical classes Wurzburg and Göttingen.
Poisson, S. D. ..	1781	Apprenticed to a surgeon, ultimately École polytechnique.
Savart, F.	1791	Medical education, then army.
Seebeck, T. J. ..	1770	Medical education.
Thénard, L. J. ..	1777	Studied pharmacy, then entered Vauquelin's laboratory.
Thompson, J. V. ..	1779	Apprentice in medicine and surgery: army surgeon.
Thomson, T.	1773	Medical studies at Edinburgh.
Wöhler, F.	1800	Medical studies at Marburg.
Wollaston, W. H. ..	1766	Medical degree, Cambridge.
Young, T.	1772	Studied for medical profession, London and Edinburgh.

TABLE VI.—Summary of Tables I–V.

Entry to a scientific profession	Number
Without higher education.	15
With higher education in non-scientific subjects	6
With higher mathematical education.	15
With scientific higher education other than medical	4
With pharmaceutical or medical education. ..	28
Total	68

It appears very clearly that the avenues that led to scientific eminence were :—

(i) The private pursuit of scientific studies by those who had no formal training.

(ii) Mathematics.

(iii) Medicine.

These three modes of education roughly correspond to the facilities available in the eighteenth century.

Self-education in science follows from the development of the taste for the subject and the availability of books that can afford a means of instruction for beginners. To understand the prevalence of self-education in science, we must therefore consider the means of popular scientific education in the eighteenth century.

The other two modes of entry follow the structure of the university courses of the time. Formal courses in mathematics were generally available, and this led to the study of its practical exemplification in mechanics, optics and astronomy. There were also formal courses in medicine, which became greatly developed in Scotland from the earlier half of the eighteenth century and in England in its closing years. These courses included instruction in chemistry and botany. The British universities also gave instruction in experimental philosophy, which covered physics and astronomy. These courses did not, however, lead to a degree or a profession, and, in fact, the attendance seems to have been small and desultory.

The same considerations seem to have applied on the continent, but here another factor emerges, namely the importance attached to military studies. The military tradition of Vauban in France, and the prestige of the wonderful succession of French mathematicians of the seventeenth and eighteenth centuries, led to a strong bias in favour of mathematical studies. It will be noted that the great scientific men who were receiving their education in the late eighteenth century were predominantly British or French: the majority of the great German scientists were born after 1800.

2. *The Popularization of Science in the Eighteenth Century.*

The tremendous import of the discoveries of the seventeenth century, especially of the unification of the cosmos by the Newtonian system, found its response in the wide popularization of science. The findings of science were conveyed to the interested public by means of encyclopædias, popularly written books and public lectures.

The first of the scientific encyclopædias written in English was John Harris's 'Lexicon Technicum,' first published in 1704. There followed throughout the eighteenth century a flood of similar and more extensive publications which were evidently widely read. Next in importance was Ephraim Chambers' 'Cyclopædia: or an Universal Dictionary of Art and Sciences' (1728), covering a wider range than Harris's work. The 'Encyclopædia Britannica or a Dictionary of Arts and Sciences, compiled by a society of gentlemen in Scotland' appeared in 1771: its special emphasis on scientific subjects is to be noted. There appeared humerous other English encyclopædic works on science, of which space forbids mention.

The same phenomenon was manifest on the continent. Thomas Corneille's 'Dictionnaire des arts et des sciences' appeared in 1694. In Germany, J. Hübner's 'Curieuses und reales Natur-Kunst-Berg-Gewerb- und Handlungs-Lexicon' appeared in 1712 and J. T. Jablonski, in 1721, published his 'Allgemeines Lexicon der Künste und Wissenschaften.' Zedler's 'Universal Lexicon' appeared in 64 volumes between 1732 and 1750. All these and others were eclipsed by the 'Encyclopédie ou dictionnaire raisonnée des sciences des arts et des mètiers par une société des gens de lettres, mis en ordre et publiée par MM. Diderot et d'Alembert,' 1751–72. This work was an admirable account of the sciences and arts. It was, moreover, an instrument of propaganda for science and its methods, and against the Church, the government and Christianity. Whatever its merits, its importance cannot be overestimated.

There were then, at the end of the eighteenth century, a number of encyclopædias dealing with special sciences, with sciences and arts, and with all knowledge, in a manner intended to inform and interest those without scientific knowledge or experience.

Numerous popular books on science appeared in the eighteenth century. In this country the most esteemed expositors of the Newtonian system were Colin Maclaurin * and James Ferguson †. The instrument-makers naturally found it to their interest to popularize scientific experiment. Thus, we find good clear works on the sciences or aspects thereof by Benjamin Martin the optician and George Adams, father and son, whose works cover a wide field in physics and astronomy and are far too numerous to record. Some popular works were addressed to "young gentlemen and ladies" and it is quite clear that science found a hearing in every type of society, from the nobleman who equipped a laboratory or observatory for the use of a man of science who enjoyed his patronage, to the apprentice seeking an education in his scanty leisure.

On the continent also many popular scientific works were published. In France the works of Voltaire and the Marquise du Châtelet were of importance in popularizing science. The former's 'Éléments de la philosophie de Newton' (1739), and the latter's translation of Newton's 'Principia' into French, published in 1756, were of importance. At a later period the influence of the works of Buffon can hardly be overestimated.

---

* An account of Sir Isaac Newton's philosophical discoveries in four books. London, 1748.

† (*a*) 'Astronomy explained upon Sir Isaac Newton's principles and made easy to those who have not studied Mathematics.' James Ferguson. London, 1756.

(*b*) 'An easy introduction to astronomy for young gentlemen and ladies. James Ferguson. 2nd Edn.' 1769.

(*c*) 'Lectures on select subjects in Mechanics, Hydrostatics, Pneumatics and Optics; with the use of the Globes; the art of Dialling; and the calculation of the Mean Times of new and full Moons and Eclipses.' James Ferguson. 1760.

Scientific periodicals date from the seventeenth century; *e. g.*, 'Philosophical Transactions'; the 'Journal des Sçavans' and the 'Acta Eruditorum.' In the latter part of the eighteenth century three tendencies are notable, the publication of new general scientific journals, of specialized scientific journals and of popular periodicals dealing with scientific matters.

Between 1770 and 1800 we may note, in the British Isles, the 'Philosophical Magazine' (1798), 'Nicholson's Journal' (1797), 'Repertory of Arts and Manufactures' (1794), 'Medical and Philosophical Commentaries' (1773), 'Botanical Magazine' (1787), 'Memoirs of the Literary and Philosophical Society of Manchester' (1785); in Germany, the 'Allgemeines Journal de Chemie,' 'Archiv für physiologie' (Leipsic, 1798), 'Der Naturforscher' (Halle, 1774), 'Journal der physik' (1790); in Switzerland, numerous biological journals, 'Magasin für die Liebhaber der Entomologie' (1778), 'Archiv der Insectengeschichte' (1781), translated into English in 1795, 'Annalen der Botanik' (1781), and others. These works were chiefly intended for men of science, but throughout the century the numerous magazines, of which the 'Gentleman's Magazine' is the best known example, frequently published scientific news in a popular form.

An important means of popular education was the hearing of courses of lectures, specially prepared for general audiences. The first of these are said to be the lectures of J. T. Desaguliers *, between 1712 and 1742. He was followed by Stephen (Triboudet) Demainbray, LL.D., who lectured at various periods between 1740 and 1780, became tutor to George III in 1754, and was afterwards first Observer of the King's Observatory at Kew. The syllabuses of some of his courses are extant. They treat Mechanicks, Motion, Hydrostaticks, Pneumaticks, Magnetism, Opticks and Astronomy, and indeed cover much the same ground as the lectures of the universities in experimental philosophy, discussed in more detail in Section 4. He endeavoured "to render the Course entertaining and improving, being particularly calculated for such GENTLEMEN and LADIES as would chuse to be acquainted with the more rational and sublimer parts of Knowledge in the most expeditious and familiar Manner." Such lectures were always copiously illustrated by demonstrations and working-models. The lecturer depended on fees from his pupils and he had the strongest reasons to make his lectures valuable and pleasant. The conditions may be gathered from a pamphlet by James Ferguson † dated 1761.

---

* The printed works of Desaguliers were of importance for popular scientific education. They include 'Physico-Mechanical Lectures, London,' 1717; 'A System of Experimental Philosophy,' London, 1719; 'A Course of Experimental Philosophy,' London, 1734.

† 'Analysis of a course of Lectures on Mechanics, Hydrostatics, Pneumatics, and Astronomy, read by James Ferguson, London MDCCLXI.'

"Any gentleman, in or near ***London***, may have this course of lectures read at his House, for Twelve Guineas, or any single Lecture for One Guinea, over and above the charge of carrying the Apparatus.

Or if twenty persons subscribe a Guinea each, they may have the whole Course read, anywhere in *London*, or within Ten miles of it.—Greater distances require larger subscriptions.

The Author teaches the Use of the Globes in a Month, attending one Hour every Day (Sunday excepted) for Two Guineas at Home . . . .

Some of these lectures, *e. g.*, those of James Arden * and James Dinwiddie †, A.M., included Chemistry.

(3) *The Teaching of Mathematics in the Eighteenth Century.*

Of this subject, the less need be said because it was a regular and established feature of University courses, and a discipline that took rank with classical and philosophical studies. Mathematical Chairs in British Universities date back to the early seventeenth or even the sixteenth centuries. Nevertheless, the standard of mathematical attainment in the British Isles was, in the period under review, much inferior to that of France, possibly owing to the general adoption on the continent of the notation of Leibniz.

In France the study of mathematics after the Revolution was strongly encouraged. In 1794 were founded the École Normale Superieure and the École Centrale des Travaux Publics, which in the next year became the École Polytechnique. The École Normale Superieure had a magnificent staff. Mathematics were taught by Lagrange, Laplace and Monge, physics by Haüy, natural history by Daubenton and chemistry by Berthollet. It does not seem, however, in its earlier years to have fostered such remarkable men as did the École Polytechnique, which was chiefly devoted to producing engineers and artillery officers. Monge, Lagrange, Legendre, Prony and Hachette, later also Poisson and Fourier, represented mathematics : Fourcroy, Vauquelin, Berthollet and Chaptal were the chemists. The standards were high and the military discipline contrasted with the general idleness that prevailed in the English Universities. Arago, in his 'Autobiography' ‡, gives a lively account of his experiences therein.

(4) *The Teaching of Physics in the Eighteenth Century.*

The sciences which are today grouped as Physics were, in the eighteenth century, treated partly as Mathematics, partly as Chemistry, but chiefly under the heading of Experimental Philosophy. The teaching of experimental philosophy began long before the period we are discussing.

---

* 'Analysis of Mr. Arden's Course of Lectures on Natural and Experimental Philosophy, by James Arden, 1782.'

† 'Syllabus of a course of Lectures on Experimental Philosophy, by James Dinwiddie, A.M., London 1790.'

‡ 'The History of my youth, an autobiography of Francis Arago,' tr. Baden Powell, London, 1855.

Thus, at Oxford, John Keill was lecturing in 1700; according to his successor, J. T. Desaguliers, he was the first to teach natural philosophy "by experiments in a mathematical manner." Lectures of this kind were given at Oxford with only short intermissions throughout the eighteenth century. We may allow ourselves a brief survey of what was taught by Thomas Hornsby *, Reader in Experimental Philosophy from 1763 to 1810. His lectures are well spoken of and a syllabus is preserved. The headings are here transcribed: suffice it to say that it is a representative specimen of this type of instruction.

A SYLLABUS OF A COURSE OF LECTURES IN EXPERIMENTAL PHILOSOPHY.

Mechaniks.

Lect. 1 Of the Properties of Bodies.
Lect. 2 Of the Laws of Motion.
Lect. 3 Of the Attraction of Cohesion.
Lect. 4 Of Magnetism.
Lect. 5 Of Electricity.
Lect. 6 Of Gravity. Of the Mechanical Powers: Of the Lever.
Lect. 7 Of the Mechanical Powers, continued.
Lect. 8 Of compound Machines: Of Friction.
Lect. 9 Of accelerated and retarded Motion, and of Pendulums.
Lect. 10 Of Central Forces.
Lect. 11 Summary View of the Solar System.
Lect. 12 Of the Figures of the Earth, and the Motion of Projectiles.

Opticks.

Lect. 13 Of Catoptricks.
Lect. 14 Of Dioptricks; of Vision.
Lect. 15 Of several Circumstances respecting Vision; of Microscopes; of Telescopes.
Lect. 16 Of the theory of Light and Colours; and the Rainbow.
Lect. 17 [No title: optical demonstration in darkened room.]

Hydrostaticks.

Lect. 18 Of the Pressure of Fluids.
Lect. 19 Of the Pressure of Fluids, continued.
Lect. 20 Of the Sinking and Floating of Solid Bodies in Fluids; of Specific Gravity.
Lect. 21 Of the Motion of Fluids.

---

* 'A syllabus of a course of Lectures in Experimental Philosophy.' No author. No date. One copy in the Museum of the History of Science is endorsed "Hornsby" in the handwriting of S. G. Rigaud.

PNEUMATICKS.

Lect. 22 Of the Air ; of the State and Limits of the Atmosphere.
Lect. 23 Of Barometers, Thermometers, and Hygroscopes.
Lect. 24 Of the Structure of the Air Pump, with a variety of Experiments to prove the Pressure of the Air.
Lect. 25 Of the Elasticity of the Air ; Structure of the Condensing Engine, etc.
Lect. 26 Miscellaneous Experiments upon Air.
Lect. 27 Different Sorts of Air.

It was illustrated by demonstrations. The apparatus apparently had to be provided by the Reader, though some colleges, *e. g.*, Oriel, had a considerable collection of " philosophical apparatus " presumably used for such lectures.

At Cambridge, Isaac Milner *, later President of Queen's College and Dean of Carlisle, read alternate courses of philosophical and chemical lectures. These continued from 1786 to 1792, and were attended with great interest. A German assistant, Hoffman, conduced to their success. Courses were also given by Antony Shepherd †, and these follow much the same lines as those of the non-academic lecturers, such as Ferguson, and of Hornsby. The lectures of Samuel Vince ‡ (1793) also follow this plan. All these lectures seem very antiquated in comparison with the active work in physics which characterized the first quarter of the nineteenth century, but they continued to be given with little change until, at least, its second decade. It will be noted that the study of heat was treated by the chemist rather than the physicist. Black's lectures ' On the effects of Heat and Mixture ' give a much more adequate account of the phenomena of heat than these lectures on experimental philosophy, where thermometry is considered as a part of pneumatics Nearly all the great physicists educated during this period came to their subject, it would seem, through mathematical studies, and not through " experimental philosophy."

(5) *Medical Teaching as a source of Scientific Education.*

The eighteenth century saw the normal mode of entry to the medical profession change from private teaching or apprenticeship to public courses at the Medical School or University. The irregular modes of entry very often produced great men—but they doubtless also produced ignorant quacks. Thus, as Flexner points out, the two Hunters, Matthew Baillie, Bright, Addison and Hodgkin began as unknown youths, in the rôle of assistants in the dead-house or the out-patient departments.

---

* ' The Life of Isaac Milner, D.D., F.R.S.,' Mary Milner, London and Cambridge, 1842.

† ' The Heads of a Course of lectures in Experimental Philosophy read at Christ College by A. Shepherd, D.D., F.R.S., and Plumian Professor of Astronomy and Experimental Philosophy, Cambridge N.D.'

‡ ' A plan of a Course of lectures on the principles of Natural Philosophy.' By the Rev. S. Vince, A.M., F.R.S., Cambridge, 1793.

Here they met with abundance of pathological and clinical material, and through that intimate acquaintance achieved their discoveries. Thereafter they became busy consultants, and research knew them no more. The medical student in a formal course listened to lectures, but had little opportunity of clinical or anatomical experience. Few schools had the opportunity to dissect half-a-dozen cadavers in the year, and since there were some 400 medical students in the largest courses, *e. g.*, that of Edinburgh, they simply looked on at demonstrations. Thus, John Bell wrote, as late as 1810.

> "In Dr. Monro's class, unless there be a fortunate succession of bloody murders, not three subjects are dissected in the year. On the remains of a subject fished up from the bottom of a tub of spirits, are demonstrated those delicate nerves which are to be avoided or divided in our operations, and these are demonstrated once at the distance of 100 feet !—nerves and arteries which the surgeon has to dissect at the peril of his patients' life."

Such a system obviously required the supplement of some kind of apprenticeship. None the less, the medical schools made admirable provision for teaching the theory and facts, as they were then known. In Great Britain, the most important schools in the eighteenth century were those of Edinburgh and Glasgow. The Edinburgh School dated from 1736. In England Guy's Hospital had a medical school of a somewhat informal character from 1723 : the London Hospital Medical School, which arose from the private medical school of Sir William Blizard and Maclaurin in 1785, is said to be the first formal medical school in England. It was followed in 1790 by St. Bartholomew's medical school, under the guidance of Abernethy. The Meath Hospital, Dublin, formed a medical school as early as 1756. At Oxford, the Medical School dates from antiquity, but the chair of clinical medicine, founded in 1780, marks a virtual refounding of the School.

In France the Revolution had a salutary effect. The *Écoles de Santé*, were founded in 1794. In 1796 the medical school of Paris was reorganized with twelve professors. Its reputation rapidly increased, and in 1799 it had 500 students.

In Germany and Austria the conditions of medical education were very variable. At Wurzburg, Heidelberg, Göttingen, and Vienna, there was adequate provision for medical teaching, but in some minor universities medical degrees could be purchased after a minimum of examination. Medical teaching normally included Chemistry, but although this was often part of the medical course, its scientific importance is such as to warrant separate consideration.

(6) *The Teaching of Chemistry.*

It is not necessary to go back very far to find the effective origins of the formal teaching of Chemistry. The subject in the eighteenth century was normally treated as an ancillary, but not an unimportant, part of medical studies, and its cultivation runs parallel to these studies. Where medical education was chiefly by apprenticeship there was little or no formal teaching of chemistry.

The great source of instruction in these subjects was the University of Edinburgh, where Medicine, Chemistry, Materia Medica and Botany were subjects of study throughout the eighteenth century. Thus Dr. James Crawford exhibited experiments in chemistry from 1713. It may be noted that no practical classes in chemistry were in existence before 1800, but some students purchased apparatus and performed the experiments at home *. A considerable advance in the teaching of Chemistry was made by William Cullen, who was appointed lecturer at Glasgow in 1746 and later moved to Edinburgh. Lyon Playfair † tells us that he

> " saw that a science like Chemistry was not to be taught by mere lectures, but that there must be a free and unreserved communication between the teacher and the taught. He cultivated the personal acquaintance of his pupils and zealously aided them to overcome those difficulties we all experience in ascending the first steps of the ladder of knowledge. He taught professors of Chemistry to act as tutors as well as prelectors . . . . "

Cullen was succeeded at Glasgow in 1756 by Joseph Black, who in 1766 was appointed to the chair of Chemistry at Edinburgh. Black's influence on the teaching of Chemistry far exceeds that of any other eighteenth-century teacher.

Black is chiefly remembered for his work on magnesia, fixed air and the alkalies, and also for his researches on latent heat. These belong to the very beginning of his career. Thereafter he did little in the way of research : he is therefore stigmatized as indolent, but in fact he gave great attention to the teaching of the subject, and therein did a great work for science. Not only were his lectures the means of educating many chemists, but they were also the pattern of the chemical lectures given at Oxford and Cambridge in the last twenty years of the century.

There are some records of his powers as a lecturer. He was an admirable manipulator ; thus Thomas Beddoes, who had been conducting a successful Chemistry course at Oxford, writes to him in 1788, in the following terms :—

> " What I find most difficult is to repeat some of those simple experiments, which in your hands are so striking and so instructive. How do you contrive to make that experiment which shows the burning of iron in dephlogisticated air ? " ‡

---

* 'History of Scottish Medicine,' Comrie, Vol. I. p. 29.

† 'A Century of Chemistry in the University of Edinburgh being the Introductory Lecture to the Course of Chemistry in 1858,' by Lyon Playfair. Edinburgh, 1853.

‡ 'Life and Letters of Joseph Black, M.D.,' Sir Wm. Ramsay, p. 89.

Robison, who edited his lectures, tells us :—

> "That while he scorned the quackery of a showman, the simplicity, neatness, and elegance with which they were performed were truly admirable."

And Brougham also :—

> "I have seen him pour boiling water or boiling acid from one vessel to another, from a vessel that had no spout into a tube, holding it at such a distance as made the stream's diameter small, and so vertical that not a drop was spilt." . . . . "The long table on which the different processes had been carried on was as clean at the end of the lecture as it had been before the apparatus was planted upon it. Not a drop of liquid, not a grain of dust, remained."

That he took the preparation for his lectures very seriously appears from a letter written to the authorities in 1789, asking for a house to be built contiguous to the chemical laboratory.

> "The Professor has, it is true, only one hour of teaching, but he must spend several hours every day in his laboratory in preparing for the experiments and operations of the next lecture or finishing those already begun: and as these operations often last ten, twelve, or twenty-four hours, or some of them several days, he is under the necessity of looking into it frequently during the day and occasionally must be there early in the morning and late at night. Nor is this sort of labour confined to the session of the college: he must have recourse to his laboratory at all times to carry on his studies, and qualify himself the better for the discharge of his duty . . . . his office is much more laborious than theirs (*the professors' of medicine*) who have only an hour of teaching daily."

Lecture-demonstrations were, of course, of the greatest importance, since the students did no practical work, and the tradition of painstaking and skilled demonstration lasted far into the nineteenth century. Faraday, for example, regarded as one of his principal duties the perfect demonstration of the phenomena treated in his lectures.

The lectures of Joseph Black have fortunately been preserved * and can be read by anyone who wishes to taste one of the principal founts of scientific education in the late eighteenth century. It is to be noted that Black did not call his work 'Lectures on Chemistry' but 'Lectures on the effect of Heat and Mixtures,' being much opposed, like Newton, to the precipitate formation of hypotheses. Black realized the imperfection of his account of these phenomena was such that it could not be regarded as forming a System of Chemistry. The style of the lectures is easy, and they are so phrased as to be comprehensible by and interesting to any intelligent person. Their arrangement seems to us to be somewhat unsystematic, with its long digressions following

---

* 'Lectures on the Elements of Chemistry, delivered in the University of Edinburgh by the late Joseph Black, M.D. . . . . published from his manuscripts by John Robison, LL.D.,' Edinburgh, 1803.

up some piece of work in which Black was peculiarly interested, but it is to be remembered that the chemistry of the time was informed by scarcely any theoretical structure and was in fact a natural history of bodies which might admit of many modes of arrangement.

Black's lectures, as edited by Robison, are in their latest form. We know that he adopted the views of Lavoisier before 1784 : but his tendency was always to give the least possible weight to theories and the greatest to the correct description of phenomena.

A good idea of what Black taught can be gathered from the headings of Robison's work which are transcribed below. It has no table of contents, the compilation of which would doubtless have revealed to the editor the anomalies of his system of headings.

Part I.

INTRODUCTORY. Definition of Chemistry. Of Chemistry in General.

GENERAL DOCTRINES OF CHEMISTRY.

PART I. GENERAL EFFECTS OF HEAT.

Introduction.	Of heat in general. Heat and Cold. Nature of Heat.
Sect. I.	Of Expansion. Of the Thermometer, Improvement of our knowledge of heat by means of the thermometer, (1) Extension of our ideas of heat, (2) Of the Distribution of heat, (3) Of the Celerity of the Communication of heat.
Sect. II.	Of fluidity.
Sect. III.	Of Vapour and Vaporization. Of Evaporation *.
Sect. IV.	Of Ignition.
Sect. V.	Of Inflammation †.

Part II.

GENERAL EFFECTS OF MIXTURE.

Introduction.	Of the theories of Chemical Mixture and Combination. Of Electric Attractions.

PART III. CHEMICAL APPARATUS.

THE CHEMICAL HISTORY OF BODIES.

Introduction.

CLASS I. SALTS.

Of Salts in General.

GENUS I. ALKALINE SALTS.

Species I	Vegetable Alkali.
Species II	Fossil Alkali.
Species III	Vegetable Alkali.

* Sect. II and III contain Black's famous work on latent heat and the last-named deals with steam engines.

† This section deals with combustion.

GENUS II. ACID SALTS.

Vitriolic Acid.

Species III (*sic*)	Nitrous and Muriatic acids. Nitrous acid.
Species III	Muriatic acid.
Species IV	Acetous acid.
Species V	Acid of tartar.
Species VI	Sedative salt.

GENUS III. OF THE COMPOUND OR NEUTRAL SALTS.

Species I	Vitriolated tartar.
Species II	Glauber's Salt.
Species III	Nitre or Saltpetre *.
Species IV	Cubic nitre †.
Species V	Digestive Salt ‡.
Species VI	Common Salt.
Species VII	Vitriolic Ammoniac.
Species VIII	Nitrous Ammoniac.
Species IX	Sal Ammoniac.
Species X	Regenerated tartar.
Species XI	The Alkali Fossile Acetatum.
Species XII	Acetous Ammoniac.
Species XIII	Tartar.
Species XIV	Borax.

Synonymous Denominations of the Salts.

CHEMICAL HISTORY.

CLASS II. EARTHS.

Introductory.

GENUS I. ALKALINE EARTHS.

Species I	Calcareous earths.
Species II	Magnesia §. Calcareous Earths combined with acids. 1. *Gypsum.* 2. Fluor or Fluat of Lime 3. Phosphat of Lime. 4. Borat of Lime.
Species III	Barytes.
Species IV	Strontites.

GENUS II. ARGILLACEOUS EARTHS.

Plastic Earths.

GENUS III. HARD STONY BODIES.

* Contains a considerable digression on oxygen, etc.

† Sodium nitrate.

‡ Potassium Chloride.

§ A long section treating of Black's work on the alkalies and fixed air, and subsequent discoveries and controversies.

GENUS IV. FUSIBLE STONES.

GENUS V. FLEXIBLE STONES.

Appendices on Precious stones, Gems, Porcelain, Quartz.

CLASS III. INFLAMMABLE SUBSTANCES.

Introductory. (On Combustion).

I. Inflammable Air *.
II. Phosphorus.
III. Sulphur.
IV. Charcoal. *Appendix.* Pyrophori.
V. Ardent Spirits.
VI. Oils. Aromatic Oils, Unctuous oils. Empyreumatic oils.
VII. Bitumens.

CLASS IV. METALLIC SUBSTANCES.

Introductory and General Account.

METALS.

Genus I Arsenic.
II Magnesium †.
III Iron ‡.
IV Mercury.
V Antimony.
VI Zinc, or Spelter.
VII Bismuth or Tinglass.
VIII Cobalt.
IX Niccolum.
X Lead.
XI Tin – Stannum.
XII Copper.
Silver and Gold.
XIII Silver.
XIV Gold.
XV Platina, or Platinum.

CLASS V. OF WATERS.

Introductory and General.

I. Experiments to discover the carbonic acid. II. Experiments to discover Alkaline Salts. III. Experiments to discover Sulphuric acid. IV. Experiments to discover the Muriatic Acid. V. Detection of Earths in Mineral Waters. VI. Detection of Sulphur.

This system of classification and, indeed, the greater part of the matter was adopted by those who set up similar courses elsewhere.

---

* Treats of composition of water.

† Under this title Black describes manganese, and describes chlorine and the chlorates.

‡ A considerable section, dealing also with the metallurgy of iron and steel.

Scotland had many admirable chemists besides Black. Thomas Charles Hope taught at Glasgow from 1787 and in 1795 succeeded Black. Hope had some 500 students in chemistry and was the first to open a laboratory for practical classes. He is remembered for his discovery of strontia and for his work on the maximum density of water. Thomas Thomson, who studied under Black before 1799 became Professor of Chemistry at Glasgow in 1818. At the Andersonian College, Dr. Andrew Ure, well-known for his Dictionary, was professor from 1804.

At Oxford the first serious lectures in chemistry since the seventeenth century were given by Martin Wall, who was appointed Public Reader in Chemistry in 1782. He has left us a very full syllabus of his course *.

In his preface he tells us that :—

> " The University during the last Course gave demonstrative Proofs of their sincere Zeal to raise a School of Medicine in this Place, by their Indulgence in Regard to the Degrees, and their liberal Assistance in the Improvement of the Laboratory."

The University did not, however, provide him with a salary, for he had to support himself by fees of two guineas a term, payable by his students.

We learn from contemporary correspondence that he had an audience of fourteen or fifteen, which dropped away at the end of the course.

The syllabus reveals that, like most unpractised lecturers, he had provided far more material than he could deliver. Thus a student wrote :—

> " It is unfortunate for us that the Doctor is tied down to a single hour, by which means his lectures are read with a rapidity that prevents much from being remembered, and anything from being taken down."

The general outline of the course resembles Black's, but the treatment of certain matters, *e. g.*, chemical apparatus and furnaces, seems to have been fuller. Moreover, the course concludes with a section on vegetable and animal chemistry.

Wall was succeeded by Thomas Beddoes, who seems to have been a lively lecturer. In the letter quoted in section 6, he tells Black that he has :—

> " the largest class in chemistry that has ever been seen at Oxford within the memory of man, in any department of knowledge, although my number cannot be put in competition with those in Edinburgh."

Neither his lectures nor even a syllabus of them survives. His popularity probably meant that he was a skilled and lively demonstrator, which is witnessed by his sending up a balloon on the 10th June, 1790, with a fuse of touch-paper arranged to set it alight after it had travelled about three miles, so as to discover whether the conflagration would give the appearance of an " igneous meteor." Beddoes found the conditions at Oxford unsatisfactory, for the fees from his class, his sole emolument, were scarcely enough to pay his expenses—for it seems

* 'A syllabus of a course of Lectures in Chemistry. Read at the Museum, Oxford.' By Martin Wall, M.D., Public Reader in Chemistry. Oxford, 1782.

that he had to find his own apparatus and materials. He was succeeded in 1784 by Robert Bourne, M.D., who has left a syllabus * of his course.

Bourne follows somewhat different lines, being more impressed with the importance of pneumatic chemistry. He starts with the general notion of attraction in natural philosophy and treats of Chemical attraction. He treats first of Light and Heat. He then gives a considerable section on aerial substances, " vital air, azotic air, inflammable air, alkaline air, fixed air, vitriolic acid air, muriatic acid air, oxygenated muriatic acid air, fluor acid air." He then treats saline substances; first the acids, then the alkalis and then the salts. He uses the new nomenclature. He then treats earthy substances from a more chemical point of view than does Wall. He continues with inflammable substances, sulphur, coal, alcohol, oils, resins, phosphorus, then goes on to the metals. He adds at the end :—

> " For the sake of conciseness, the Reader in Chemistry has not specified the experiments, which will be shewn in his course of Lectures. But he thinks it right to observe, in a general manner, that as many experiments will be shown, and as many processes exhibited as the time will permit; in short that the Course will be practical rather than theoretical."

In 1803 John Kidd was appointed to the newly founded Aldrichian Professorship of Chemistry at Oxford. In 1808, he published a syllabus of a course of lectures which cover very much the same ground as those of Bourne. But the influence of the new way of thinking is manifest. Gases are no longer treated together as varieties of air, but are arranged according to their constituents. Light and Heat are still included in the course, but it seems that the discussion given to them was not so full.

Kidd scarcely kept pace with the rapid progress of chemistry. In 1818 he took exception to a statement of Brande in the supplement to the fourth and fifth editions of the 'Encyclopædia Britannica,' that " excepting in the Schools of London and Edinburgh, Chemistry, as a branch of education, is either neglected or, what is perhaps worse, superficially and imperfectly taught." Kidd rebuts this charge by printing the syllabus of his lectures, and incidentally gives his views on scientific education. He does not consider the purpose of such teaching to be primarily an introduction to a scientific career, but states :—

> " It is evident, then to those who reflect on the subject, that the whole tenor of an academical education, so far at least as intellectual endowments are concerned, regards the general improvement of its members, rather than their qualification for any particular profession : and hence the trite objection so often even now brought forward, that the Physical and Experimental Sciences are here neglected, can only proceed from want of candour or of information. For a candid and enlightened mind would readily allow, that though the discipline of

* 'A Syllabus of a Course of Chemical Lectures read at the Museum Oxford in seventeen hundred ninety four.' (No author's name.)

† Syllabus of a Course of Lectures on Chemistry (by J. Kidd, M.A.). Oxford, 1808.

Classical and Mathematical studies is well calculated to form the groundwork of excellence in the Physical and Experimental Sciences, the converse of this is by no means true ; witness the deficiency, with respect to taste and reasoning, in the literary productions of individuals, whose fame in other points deservedly ranks high in the scientific and professional world.

The Physical and Experimental Sciences then are not neglected in this place. They are not cultivated indeed to the same extent as in some other schools ; but they are cultivated so far as is compatible with the views of a system of general education : and hence the object of the Lectures in the several branches of those sciences is, rather to present a liberal illustration of their principles and practical application, than to run into the minutiæ of a technical, or even a philosophical detail of facts. These branches of science, in this place at least, may be considered with reference to Divinity, Classics, and Mathematics, in the same light as the supernumerary war-horses of Homer's chariots : which were destined to assist, but not to regulate the progress of their nobler fellow coursers . . . .

It has been the means, that is, of elevating to the title of Philosophers a host of individuals, whose talents were just equal to that species of inductive reasoning, the nature of which has been of late years so egregiously mistaken, and its importance so absurdly vaunted. That man, in truth, must be possessed of but ordinary abilities, who cannot draw a general conclusion from a number of analogous facts continually passing before his eyes ; while, after all, it must be genius alone that can penetrate beyond the limits which apparently confine it, and connect at once the distant or hidden links in a chain of philosophic reasoning." (He argues that the powers of such a genius will be improved and exercised by a general education, though some are found with talents that supersede the necessity of any education.)

The Professorship of Chemistry at Cambridge was founded in 1713, but the first holder of it we may notice is that remarkable character Richard Watson, later Bishop of Llandaff. In his memoirs * he tells us with noble candour how he was appointed :—

> "In October 1764, I was made moderator for Christ's College. On the 19th of the following November, on the death of Dr. Hadley, I was unanimously elected by the Senate, assembled in full congregation, Professor of Chemistry. An eminent physician in London had expressed a wish to succeed Dr. Hadley, but on my signifying to him that it was my intention to read chemical lectures in the University, he declined the contest. At the time this honour was conferred upon me, I knew nothing at all of Chemistry, had never read a syllable on the subject ; nor seen a single experiment in it ; but I was tired with mathematics and natural philosophy, and the *vehementissima gloriae cupido* stimulated me to try my strength in a new pursuit, and the kindness of the University (it was always very kind to me) animated me to very extraordinary exertions. I sent immediately after my election for an operator to Paris ; I buried myself as it were in my laboratory, at least as much as my other avocations would permit ; and in fourteen months from my election, I read a course of chemical lectures to a very full audience, consisting of persons of all ages and degrees, in the University."

---

* 'Anecdotes of the life of R. W. written by himself at different intervals, and revised in 1814.' Richard Watson. London, 1817.

We may gain some idea of Bishop Watson's lectures from his volumes of Chemical Essays *. They are pleasantly and lightly written accounts of interesting matters concerned with chemistry, but make little or no attempt to grapple with the unsolved problems of the subject. It is indeed such a work as a highly educated and intelligent man could compile in a year's reading, but it smells rather of the lamp than the laboratory. The experiments which he narrates as his own are chiefly concerned with manufacturing processes rather than with the advancement of pure science. He tells us † :—

> "Chemistry is cultivated abroad by persons of the first Rank, Fortune and Ability; they find in it a never failing source of honourable amusement for their private hours; and as public men, they consider its cultivation as one of the most certain means of bringing to perfection, the manufactures of their country ‡."

Watson was followed by a much greater man, Francis John Hyde Wollaston. We possess the syllabus of his lectures ‡. Their plan evidently derives from Black, but Wollaston uses the new nomenclature, and introduces a number of substances recently discovered.

A course of an entirely new type was initiated by William Farish, and is fully summarized in a printed syllabus §. This interesting course begins with practical mining and metallurgy. Under each metal he treats of practical applications, *e. g.*, tool-making under iron, engraving under copper. He treats of coal-mining; and of marble, slate, clay, flints, pottery, glass, optical glasses, vitriol, salt, nitre, gunpowder, sulphuric acid, alum, &c. He further treats of animal and vegetable substances, covering agriculture, charcoal-burning, woodworking, the industries connected with oil, soap, candles, whale-fishery, tannery, sugar-refining, textile industries, printing and dyeing. He then treats the construction of machines,—windmills, water-mills, steam-engines, various mechanical powers, lathes, cranes, jacks. He further treats water-works, *i. e.*, water-supply, pumps, canals, locks, docks, ship-construction.

If the lectures corresponded to the promise of the syllabus, they must have been of the first interest and are an eloquent witness of the effect of the industrial revolution. Something of the same practical character is shown in Davy's lectures ||.

Davy begins, like Bourne, with a general discussion of the chemical powers. He then discusses first "Undecompounded substances, *i. e.*,

---

* 'Chemical Essays by R. Watson, D.D., F.R.S., and Regius Professor of Divinity in the University of Cambridge.' London, 1781.

† *Ibid.* Dedication, p. ii.

‡ 'Plan of a course of Chemical Lectures,' by Francis John Hyde Wollaston, M.A., F.R.S., Jacksonian Professor in the University of Cambridge. Cambridge, 1794.

§ 'A Plan of a Course of Lectures on Arts and Manufacturers most particularly such as relate to Chemistry.' By William Farish, M.A., Professor of Chemistry in the University of Cambridge, 1796.

|| 'Syllabus of a course of Lectures on Chemistry delivered at the Royal Institution of Great Britain' (by Sir Humphry Davy). London, 1802.

elements, then binary compounds, then bodies composed of more than two simple substances. The treatment is recognizably that of the modern text-book and contrasts remarkably with that of Wall, twenty years earlier. He then treats the general phenomena of chemical action; goes on to the Chemistry of Imponderable substances, Heat or Caloric, Light, Electrical Influence, Galvanism; then treats of the Chemistry of the Arts, Agriculture, Tanning, Bleaching, Dyeing, Metallurgy, Glass, Porcelain, Food and Drink, and the Management of Heat and Light artificially produced."

(7) *Summary.*

Surveying the position of the teaching of the physical sciences at the end of the eighteenth century, we note first a general and living interest in science, both as a means of understanding the material world and of developing the practical arts which were already beginning to revolutionize human existence. That general interest had led to the means of satisfying it through scientific works—encyclopædias, books and periodicals—and through courses of private lectures.

Almost the only scientific professions were those of the medical man, and to a less extent the engineer, mechanical or military. We find, therefore, the medical schools teaching such departments of science as were ancillary to the making of physicians and surgeons, and the military academies giving instruction in mathematics and mechanics. The Universities, whether in England or abroad, if they taught science at all, did so in a somewhat perfunctory manner, with the object of giving those who were pursuing other studies the advantage of a superficial acquaintance with the scientific discoveries of the age. None the less, the desire for better instruction was already present by 1800, and gradually came to fruition throughout the succeeding century.